Studies in Computational Intelligence

Volume 773

Series editor

Janusz Kacprzyk, Polish Academy of Sciences, Warsaw, Poland
e-mail: kacprzyk@ibspan.waw.pl

The series "Studies in Computational Intelligence" (SCI) publishes new developments and advances in the various areas of computational intelligence—quickly and with a high quality. The intent is to cover the theory, applications, and design methods of computational intelligence, as embedded in the fields of engineering, computer science, physics and life sciences, as well as the methodologies behind them. The series contains monographs, lecture notes and edited volumes in computational intelligence spanning the areas of neural networks, connectionist systems, genetic algorithms, evolutionary computation, artificial intelligence, cellular automata, self-organizing systems, soft computing, fuzzy systems, and hybrid intelligent systems. Of particular value to both the contributors and the readership are the short publication timeframe and the world-wide distribution, which enable both wide and rapid dissemination of research output.

More information about this series at http://www.springer.com/series/7092

Andrew Pownuk · Vladik Kreinovich

Combining Interval, Probabilistic, and Other Types of Uncertainty in Engineering Applications

 Springer

Andrew Pownuk
Computational Science Program
University of Texas at El Paso
El Paso, TX
USA

Vladik Kreinovich
Computational Science Program
University of Texas at El Paso
El Paso, TX
USA

ISSN 1860-949X   ISSN 1860-9503   (electronic)
Studies in Computational Intelligence
ISBN 978-3-030-08158-4   ISBN 978-3-319-91026-0   (eBook)
https://doi.org/10.1007/978-3-319-91026-0

Printed on acid-free paper

This Springer imprint is published by the registered company Springer International Publishing AG
part of Springer Nature
The registered company address is: Gewerbestrasse 11, 6330 Cham, Switzerland

Contents

1 Introduction 1
  1.1 Need for Data Processing 1
  1.2 Need to Take Uncertainty Into Account When Processing
       Data 2
  1.3 How to Gauge the Accuracy of the Estimates $\widetilde{x}_i$ 2
  1.4 Measurement and Estimation Inaccuracies Are Usually Small 3
  1.5 How to Estimate Partial Derivatives c_i 4
  1.6 Existing Methods for Computing the Probabilistic
       Uncertainty: Linearization Case 5
  1.7 Existing Methods for Computing the Interval Range:
       Linearization Case 6
  1.8 Existing Methods for Estimating Fuzzy Uncertainty:
       Linearization Case 10
  1.9 Open Problems and What We Do in This Book 11

2 How to Get More Accurate Estimates 13
  2.1 In System Identification, Interval Estimates Can Lead
       to Much Better Accuracy than the Traditional Statistical
       Ones: General Algorithm and Case Study 13
      2.1.1 Formulation of the Problem 14
      2.1.2 System Identification Under Interval Uncertainty:
          General Algorithm 16
      2.1.3 System Identification Under Interval Uncertainty:
          Simplest Case of Linear Dependence on One Variable 20
      2.1.4 Case Study 23
      2.1.5 Conclusions 24
  2.2 Which Value $\widetilde{x}$ Best Represents a Sample $x_1, \ldots, x_n$:
       Utility-Based Approach Under Interval Uncertainty 25
      2.2.1 Which Value $\widetilde{x}$ Best Represents a Sample $x_1, \ldots, x_n$:
          Case of Exact Estimates 25

       2.2.2 Case of Interval Uncertainty: Formulation
             of the Problem 27
       2.2.3 Analysis of the Problem........................ 27
       2.2.4 Resulting Algorithm........................... 29
  2.3 How to Get More Accurate Estimates – By Properly Taking
      Model Inaccuracy into Account......................... 30
       2.3.1 What if We Take into Account Model Inaccuracy 30
       2.3.2 How to Get Better Estimates 32
       2.3.3 Can We Further Improve the Accuracy? 38
  2.4 How to Gauge the Accuracy of Fuzzy Control
      Recommendations: A Simple Idea....................... 40
       2.4.1 Formulation of the Problem 40
       2.4.2 Main Idea 41
       2.4.3 But What Should We Do in the Interval-Valued
             Fuzzy Case? 42

3 How to Speed Up Computations 45
  3.1 How to Speed Up Processing of Fuzzy Data 45
       3.1.1 Cases for Which a Speedup Is Possible: A Description ... 46
       3.1.2 Main Idea of This Section 48
  3.2 Fuzzy Data Processing Beyond min t-Norm................. 49
       3.2.1 Need for Fuzzy Data Processing.................. 50
       3.2.2 Possibility of Linearization 54
       3.2.3 Efficient Fuzzy Data Processing for the min t-Norm:
             Reminder 57
       3.2.4 Efficient Fuzzy Data Processing Beyond min t-Norm:
             the Main Result of This Section 59
       3.2.5 Resulting Linear Time Algorithm for Fuzzy Data
             Processing Beyond min t-Norm 61
       3.2.6 Conclusions............................... 62
  3.3 How to Speed Up Processing of Probabilistic Data............ 63
       3.3.1 Need to Speed Up Data Processing Under
             Uncertainty: Formulation of the Problem.............. 63
       3.3.2 Analysis of the Problem and Our Idea 65
       3.3.3 How to Approximate 67
       3.3.4 Resulting Algorithm........................... 69
       3.3.5 Numerical Example 70
       3.3.6 Non-uniform Distribution of α_j is Better 71
  3.4 Hypothetical Quantum-Related Negative Probabilities Can
      Speed Up Uncertainty Propagation Algorithms............... 76
       3.4.1 Introduction............................... 76
       3.4.2 Uncertainty Propagation: Reminder and Precise
             Formulation of the Problem 77

3.4.3   Existing Algorithms for Uncertainty Propagation
        and Their Limitations.   82
3.4.4   Analysis of the Problem and the Resulting
        Negative-Probability-Based Fast Algorithm
        for Uncertainty Quantification   87

**4   Towards a Better Understandability of Uncertainty-Estimating
     Algorithms**   97
4.1   Case of Interval Uncertainty: Practical Need for Algebraic
      (Equality-Type) Solutions of Interval Equations
      and for Extended-Zero Solutions   97
      4.1.1   Practical Need for Solving Interval Systems
              of Equations: What Is Known   98
      4.1.2   Remaining Problem of How to Find the Set A
              Naturally Leads to Algebraic (Equality-Type)
              Solutions to Interval System of Equations   102
      4.1.3   What if the Interval System of Equations Does
              not Have an Algebraic (Equality-Type) Solution:
              A Justification for Enhanced-Zero Solutions   103
4.2   Case of Interval Uncertainty: Explaining the Need
      for Non-realistic Monte-Carlo Simulations   106
      4.2.1   Problem: The Existing Monte-Carlo Method
              for Interval Uncertainty Is not Realistic.   106
      4.2.2   Proof That Realistic Interval Monte-Carlo Techniques
              Are not Possible: Case of Independent Variables   107
      4.2.3   Proof That Realistic Interval Monte-Carlo Techniques
              Are not Possible: General Case   108
      4.2.4   Why Cauchy Distribution.   114
4.3   Case of Probabilistic Uncertainty: Why Mixture
      of Probability Distributions?   115
      4.3.1   Formulation of the Problem   116
      4.3.2   Main Result of This Section   116
4.4   Case of Fuzzy Uncertainty: Every Sufficiently Complex
      Logic Is Multi-valued Already.   119
      4.4.1   Formulation of the Problem: Bridging the Gap Between
              Fuzzy Logic and the Traditional 2-Valued Fuzzy
              Logic   119
      4.4.2   Fuzzy Logic – The Way It Is Typically Viewed
              as Drastically Different from the Traditional
              2-Valued Logic   121
      4.4.3   There Is Already Multi-valuedness in the Traditional
              2-Valued Fuzzy Logic: Known Results   121
      4.4.4   Application of 2-Valued Logic to Expert Knowledge
              Naturally Leads to a New Aspect of Multi-valuedness –
              An Aspect Similar to Fuzzy   122

4.5 General Case: Explaining Ubiquity of Linear Models 125
   4.5.1 Formulation of the Problem 125
   4.5.2 Analysis of the Problem: What Are Reasonable
         Property of an Interpolation 126
   4.5.3 Main Result of This Section 128
4.6 General Case: Non-linear Effects 132
   4.6.1 Compaction Meter Value (CMV) – An Empirical
         Measure of Pavement Stiffness 132
   4.6.2 A Possible Theoretical Explanation of an Empirical
         Correlation Between CMV and Stiffness 133

5 How General Can We Go: What Is Computable and What
   Is not 137
   5.1 Formulation of the Problem 137
   5.2 What Is Computable: A Brief Reminder 138
   5.3 What We Need to Compute: A Even Briefer Reminder 142
   5.4 Simplest Case: A Single Random Variable 142
   5.5 What if We Only Have Partial Information About
       the Probability Distribution? 146
   5.6 What to Do in a General (not Necessarily 1-D) Case 148
   5.7 Proofs 150
   5.8 Conclusions 154

6 Decision Making Under Uncertainty 157
   6.1 Towards Decision Making Under Interval Uncertainty 157
      6.1.1 Formulation of the Practical Problem 158
      6.1.2 Formulation of the Problem in Precise Terms 158
      6.1.3 Analysis of the Problem 159
      6.1.4 Explaining Why, In General, We Have $f_{x_i} \neq 0$ 161
      6.1.5 Analysis of the Problem (Continued) 162
      6.1.6 Solution to the Problem 166
   6.2 What Decision to Make In a Conflict Situation Under Interval
       Uncertainty: Efficient Algorithms for the Hurwicz Approach 167
      6.2.1 Conflict Situations Under Interval Uncertainty:
            Formulation of the Problem and What Is Known
            So Far 168
      6.2.2 Conflict Situation Under Hurwicz-Type Interval
            Uncertainty: Analysis of the Problem 172
      6.2.3 Algorithm for Solving Conflict Situation Under
            Hurwicz-Type Interval Uncertainty 174
      6.2.4 Conclusion 176

6.3   Decision Making Under Probabilistic Uncertainty: Why
        Unexpectedly Positive Experiences Make Decision Makers
        More Optimistic   176
        6.3.1   Formulation of the Problem   177
        6.3.2   Formulating the Problem in Precise Terms   177
        6.3.3   Towards the Desired Explanation   178
6.4   Decision Making Under General Uncertainty   180
        6.4.1   Formulation of the Problem   180
        6.4.2   Analysis of the Problem   182
        6.4.3   Conclusions   187

7   Conclusions   191

References   193

Index   199

Abstract

In many practical applications, we process measurement results and expert estimates. The measurements and expert estimates are never absolutely accurate, and their results are slightly different from the actual (unknown) values of the corresponding quantities. It is therefore desirable to analyze how these measurement and estimation inaccuracies affect the results of data processing.

There exist numerous methods for estimating the accuracy of the results of data processing under different models of measurement and estimation inaccuracies: probabilistic, interval, and fuzzy. To be useful in engineering applications, these methods should provide accurate estimate for the resulting uncertainty, should not take too much computation time, should be understandable to engineers, and should be sufficiently general to cover all kinds of uncertainty.

In this book, on several case studies, we show how we can achieve these four objectives. We show that we can get more accurate estimates, for example, by properly taking model inaccuracy into account. We show that we can speed up computations, e.g., by processing different types of uncertainty separately. We show that we can make uncertainty-estimating algorithms more understandable, e.g., by explaining the need for non-realistic Monte Carlo simulations. We also analyze how to make decisions under uncertainty and how general uncertainty-estimating algorithms can be.

Chapter 1
Introduction

1.1 Need for Data Processing

One of the main objectives of science is to predict future values of physical quantities. For example:

- in meteorology, we need to predict future weather;
- in airplane control, we need to predict the location and the velocity of the plane under current control, etc.

To make these predictions, we need to know how each of these future values y depends on the current values $x_1, \ldots, x_n$ of the related quantities, i.e., we need to know an algorithm $y = f(x_1, \ldots, x_n)$ that relates y to x_i.

Once we find this information, we can then use the results $\widetilde{x}_1, \ldots, \widetilde{x}_n$ of measuring the quantities x_i to compute an estimate $\widetilde{y} = f(\widetilde{x}_1, \ldots, \widetilde{x}_n)$ for the desired future value y. In these terms, the prediction consists of two stages:

- first, we measure or estimate the values of the quantities $x_1, \ldots, x_n$;
- then, we use the results $\widetilde{x}_i$ of measurement or estimation to compute an estimate $\widetilde{y}$ of the desired future value y as

$$\widetilde{y} = f(\widetilde{x}_1, \ldots, \widetilde{x}_n). \tag{1.1.1}$$

This computation is an important case of *data processing*.

For example, to predict tomorrow's temperature in El Paso y, we need to know today's temperature, wind speed and direction, and humidity in different locations inside El Paso and near El Paso; these values $x_1, \ldots, x_n$ are what we can use for this prediction. We can then use an appropriate method for solving the corresponding partial differential equation as the desired prediction algorithm $y = f(x_1, \ldots, x_n)$.

The weather example shows that the corresponding prediction algorithms can be very complicated; thus, we need to use high-performance computers for this data processing.

© Springer International Publishing AG, part of Springer Nature 2018

A. Pownuk and V. Kreinovich, *Combining Interval, Probabilistic, and Other Types of Uncertainty in Engineering Applications*, Studies in Computational Intelligence 773, https://doi.org/10.1007/978-3-319-91026-0_1

1

Other situations when we need data processing come from the fact that we also want to know the current state of the world, i.e., we want to know the current values of all the quantities that describe this state. Some of these quantities – like temperature in El Paso – we can measure directly. Other quantities, such as the temperature or the density deep inside the Earth, are difficult or even impossible to measure directly. To find the values of each such difficult-to-measure quantity y, a natural idea is to find related easier-to-measure quantities $x_1, \ldots, x_n$ that are related to the desired quantity y by a known dependence $y = f(x_1, \ldots, x_n)$, and then use the results $\widetilde{x}_1, \ldots, \widetilde{x}_n$ of measuring x_i to compute an estimate $\widetilde{y} = f(\widetilde{x}_1, \ldots, \widetilde{x}_n)$ for y.

1.2   Need to Take Uncertainty Into Account When Processing Data

In general, data processing means applying some algorithm $f(x_1, \ldots, x_n)$ to the values of the quantities $x_1, \ldots, x_n$, resulting in a value $y = f(x_1, \ldots, x_n)$.

Values x_i usually come from measurements. Measurement are never absolutely accurate; the measurement result $\widetilde{x}_i$ is, in general, different from the actual (unknown) value x_i of the corresponding quantity: $\Delta x_i \stackrel{\text{def}}{=} \widetilde{x}_i - x_i \neq 0$; see, e.g., [82].

Because of the this, the computed value $\widetilde{y} = f(\widetilde{x}_1, \ldots, \widetilde{x}_n)$ is, in general, different from the ideal value $y = f(x_1, \ldots, x_n)$.

It is therefore desirable to estimate the accuracy $\Delta y \stackrel{\text{def}}{=} \widetilde{y} - y$. To estimate Δy, we need to have some information about the measurement errors Δx_i.

1.3   How to Gauge the Accuracy of the Estimates $\widetilde{x}_i$

In this book, we consider two types of estimates: measurements and expert estimates.

For *measurements*, we usually know the upper bound Δ_i on the absolute value of the measurement error $\Delta x_i \stackrel{\text{def}}{=} \widetilde{x}_i - x_i$; see, e.g., [82]. This upper bound is usually provided by the manufacturer of the measurement instrument. The existence of such an upper bound comes from the very nature of measurement: if no upper bound is guaranteed, this means that whatever result the "measuring instrument" produces, the actual value can be any number from $-\infty$ to $+\infty$; this would be not a measurement result, it would be a wild guess.

Once we know the upper bound Δ_i for which $|\Delta x_i| \leq \Delta_i$, and we know the measurement result $\widetilde{x}_i$, then we know that the actual value x_i is located in the interval

$$[\widetilde{x}_i - \Delta_i, \widetilde{x}_i + \Delta_i].$$

Different values x_i from these intervals can lead, in general, to different values of $y = f(x_1, \ldots, x_n)$. Our goal is then to find the range **y** of all possible values of y:

$$\mathbf{y} = \{f(x_1, \ldots, x_n) : x_1 \in \mathbf{x}_1, \ldots, x_n \in \mathbf{x}_n\}. \tag{1.3.1}$$

The problem of computing this range $\mathbf{y}$ is one of the main problems of *interval computations*; see, e.g., [26, 58, 82].

Often, in addition to the upper bound Δ_i on the measurement error, we also know the probabilities of different values $\Delta x_i \in [-\Delta_i, \Delta_i]$; see, e.g., [30, 31, 82].

To gauge the accuracy of (fuzzy) *estimates*, it is reasonable to use fuzzy techniques, techniques specifically designed to describe imprecise ("fuzzy") expert estimates in precise computer-understandable terms; see, e.g., [28, 61, 108]. In these techniques, the uncertainty of each estimate is described by a *membership function* $\mu_i(x_i)$ that describes, for all possible real numbers x_i, the degree to which the expert believes that this number is a possible value of the corresponding quantity.

1.4 Measurement and Estimation Inaccuracies Are Usually Small

In many practical situations, the measurement and estimation inaccuracies Δx_i are relatively small, so that we can safely ignore terms which are quadratic (or of higher order) in terms of Δx_i [82]. We can use this fact to simplify the expression for the inaccuracy $\Delta y \stackrel{\text{def}}{=} \widetilde{y} - y$.

Here, by definition of data processing, $\widetilde{y} = f(\widetilde{x}_1, \ldots, \widetilde{x}_n)$. Thus,

$$\Delta y = \widetilde{y} - y = f(\widetilde{x}_1, \ldots, \widetilde{x}_n) - f(x_1, \ldots, x_n). \tag{1.4.1}$$

From the definition of the measurement uncertainty Δx_i, we conclude that $x_i = \widetilde{x}_i - \Delta x_i$. Substituting this expression into the above formula (1.4.1) for Δy, we conclude that

$$\Delta y = \widetilde{y} - y = f(\widetilde{x}_1, \ldots, \widetilde{x}_n) - f(\widetilde{x}_1 - \Delta x_1, \ldots, \widetilde{x}_n - \Delta x_n). \tag{1.4.2}$$

Expanding the expression $f(\widetilde{x}_1 - \Delta x_1, \ldots, \widetilde{x}_n - \Delta x_n)$ in Taylor series in terms of small values Δx_i, and using the fact that terms quadratic in Δx_i can be safely ignored, we conclude that

$$f(\widetilde{x}_1 - \Delta x_1, \ldots, \widetilde{x}_n - \Delta x_n) = f(\widetilde{x}_1, \ldots, \widetilde{x}_n) - \sum_{i=1}^{n} c_i \cdot \Delta x_i, \tag{1.4.3}$$

where we denoted

$$c_i \stackrel{\text{def}}{=} \frac{\partial f}{\partial x_i}. \tag{1.4.4}$$

Substituting the expression (1.4.3) into the formula (1.4.2) and cancelling out the terms $+f(\widetilde{x}_1, \ldots, \widetilde{x}_n)$ and $-f(\widetilde{x}_1, \ldots, \widetilde{x}_n)$, we conclude that

$$\Delta y = \sum_{i=1}^{n} c_i \cdot \Delta x_i. \tag{1.4.5}$$

This is the main formula used to estimate the accuracy of estimating the desired quantity y.

1.5 How to Estimate Partial Derivatives c_i

Case of analytical differentiation. In some cases, we have explicit expressions – or efficient algorithms – for the partial derivatives (1.4.4).

Numerical differentiation: idea. In many practical situations, we do not have algorithms for computing the derivatives c_i. This happens, e.g., when we use proprietary software in our computations – in this case, we cannot use neither formula for differentiation, nor automatical differentiation tools.

In such situations, we can use the fact that we are under the linearization assumption that thus, that for each i and $h_i \neq 0$, we have

$$f(\widetilde{x}_1, \ldots, \widetilde{x}_{i-1}, \widetilde{x}_i + h_i, \widetilde{x}_{i+1}, \ldots, \widetilde{x}_n) \approx f(\widetilde{x}_1, \ldots, \widetilde{x}_{i-1}, \widetilde{x}_i, \widetilde{x}_{i+1}, \ldots, \widetilde{x}_n) + h_i \cdot c_i. \tag{1.5.1}$$

If we move the term $\widetilde{y} = f(\widetilde{x}_1, \ldots, \widetilde{x}_{i-1}, \widetilde{x}_i, \widetilde{x}_{i+1}, \ldots, \widetilde{x}_n)$ to the left-hand side of the formula (1.5.1) and divide both sides of the resulting approximate equality by h_i, we conclude that

$$c_i \approx \frac{f(\widetilde{x}_1, \ldots, \widetilde{x}_{i-1}, \widetilde{x}_i + h_i, \widetilde{x}_{i+1}, \ldots, \widetilde{x}_n) - \widetilde{y}}{h_i}. \tag{1.5.2}$$

This is a known formula for numerical differentiation.

Numerical differentiation: algorithm. We select some values $h_i \neq 0$. Then, we compute the values

$$c_i = \frac{f(\widetilde{x}_1, \ldots, \widetilde{x}_{i-1}, \widetilde{x}_i + h_i, \widetilde{x}_{i+1}, \ldots, \widetilde{x}_n) - \widetilde{y}}{h_i}. \tag{1.5.3}$$

Numerical differentiation: computation time. The above algorithm contains $n + 1$ calls to the original data processing algorithm f:

- one call to compute $\widetilde{y}$ and
- n calls to compute n partial derivatives $c_1, \ldots, c_n$.

As we have mentioned earlier, the data processing algorithm f itself can be very time-consuming. The same weather prediction example shows that the number n of input variables can also be large, in hundreds or even thousands. As a result, the computation time needed for the numerical differentiation method can be very large.

1.6   Existing Methods for Computing the Probabilistic Uncertainty: Linearization Case

Analysis of the problem. In the probabilistic approach, it is assumed that we know the probability distributions of the measurement errors Δx_i, Usually, it is assumed that the measurement errors Δx_i are independent, and that each of them is normally distributed with 0 mean and known standard deviation σ_i.

Under these assumptions, the linear combination $\Delta y = \sum_{i=1}^{n} c_i \cdot \Delta x_i$ is also normally distributed, with 0 mean and variance

$$\sigma^2 = \sum_{i=1}^{n} c_i^2 \cdot \sigma_i^2. \tag{1.6.1}$$

First method: numerical differentiation. In principle, we can estimate all the partial derivatives c_i and then apply the above formula.

Need for a faster method. As we have mentioned, the numerical differentiation method takes too long time. It is therefore desirable to have a faster method for computing σ.

Monte-Carlo method. Such a faster method is known: it is the method of Monte-Carlo simulations. In this method, to find the desired distribution for Δy, we several times $k = 1, \ldots, N$, do the following:

- we simulate n variables $\Delta x_i^{(k)}$ according to the corresponding probability distribution $\rho_i(\Delta x_i)$ (usually, normal);
- then we simulate $x_i^{(k)} = \tilde{x}_i - \Delta x_i^{(k)}$ for each i;
- we apply the data processing algorithm $f(x_1, \ldots, x_n)$ to the simulated values, resulting in $y^{(k)} = f(x_1^{(k)}, \ldots, x_n^{(k)})$;
- finally, we compute $\Delta y^{(k)} = \tilde{y} - y^{(k)}$.

One can easily check that these differences $\Delta y^{(k)}$ have the same distribution as Δy. So, we can determine the desired probability distribution from the sample

$$\Delta y^{(1)}, \ldots, \Delta y^{(N)}.$$

1.7 Existing Methods for Computing the Interval Range: Linearization Case

Analysis of the problem. The expression (1.4.5) for Δy attains its largest value when each of the terms $c_i \cdot \Delta x_i$ attains its largest possible value.

Each of these terms is a linear function of Δx_i on the interval $[-\Delta_i, \Delta_i]$. When $c_i \geq 0$, this linear function is increasing and thus, it attains its largest possible value when Δx_i is the largest, i.e., when $\Delta x_i = \Delta_i$. The corresponding value of the term $c_i \cdot \Delta x_i$ is $c_i \cdot \Delta_i$.

When $c_i < 0$, the linear function $c_i \cdot \Delta x_i$ is decreasing and thus, it attains its largest possible value when Δx_i is the smallest, i.e., when $\Delta x_i = -\Delta_i$. The corresponding value of the term $c_i \cdot \Delta x_i$ is $c_i \cdot (-\Delta_i) = (-c_i) \cdot \Delta_i$.

In both cases, the largest possible value of each term $c_i \cdot \Delta x_i$ is equal to $|c_i| \cdot \Delta_i$. Thus, the largest possible value Δ of the sum (1.4.5) is equal to

$$\Delta = \sum_{i=1}^{n} |c_i| \cdot \Delta_i. \tag{1.7.1}$$

Similarly, one can show that the smallest possible value of the sum (1.4.5) is equal to $-\Delta$. Thus, the range of possible values of Δy is the interval $[-\Delta, \Delta]$, and the range **y** of possible values of $y = \widetilde{y} - \Delta y$ is equal to

$$\mathbf{y} = [\widetilde{y} - \Delta, \widetilde{y} + \Delta]. \tag{1.7.2}$$

Thus, once we have the result $\widetilde{y} = f(\widetilde{x}_1, \ldots, \widetilde{x}_n)$ of data processing, to compute the desired range **y**, it is sufficient to be able to compute the value Δ.

First method: numerical differentiation. In principle, we can estimate all the partial derivatives c_i and then apply the above formula.

In particular, for $h_i = \Delta_i$, we get

$$\Delta = \sum_{i=1}^{n} |f(\widetilde{x}_1, \ldots, \widetilde{x}_{i-1}, \widetilde{x}_i + \Delta_i, \widetilde{x}_{i+1}, \ldots, \widetilde{x}_n) - \widetilde{y}|. \tag{1.7.3}$$

Need for a faster method. As we have mentioned, the numerical differentiation method takes too long time. It is therefore desirable to have a faster method for computing σ.

Monte-Carlo method. Such a faster method is indeed known; see, e.g., [34]. This method is based on using Cauchy distribution, with the probability density function

$$\rho_\Delta(x) = \frac{\Delta}{\pi} \cdot \frac{1}{1 + \dfrac{x^2}{\Delta^2}}. \tag{1.7.4}$$

Specifically, there is a known result about this distribution: that

- when we have several independent random variables Δx_i distributed according to Cauchy distribution with parameter Δ_i,
- then their linear combination $\sum_{i=1}^{n} c_i \cdot \Delta x_i$ is also Cauchy distributed, with parameter

$$\Delta = \sum_{i=1}^{n} |c_i| \cdot \Delta_i.$$

This is exactly the desired formula (1.7.1). Thus, we can find Δ as follows:

- first, we several times simulate the inputs $\Delta x_i^{(k)}$ according to the Cauchy distribution;
- then, we plug in the corresponding simulated values $x_i^{(k)} = \widetilde{x}_i - \Delta x_i^{(k)}$ into the data processing algorithm $f(x_1, \ldots, x_n)$, producing the values $y^{(k)} = f(x_1^{(k)}, \ldots, x_n^{(k)})$;
- then, the differences $\Delta y^{(k)} = \widetilde{y} - y^{(k)}$ are also Cauchy distributed, with the desired parameter Δ.

The desired value Δ can then be determined, e.g., by using the Maximum Likelihood method, i.e., from the condition that

$$L \stackrel{\text{def}}{=} \prod_{k=1}^{N} \rho_\Delta(\Delta y^{(k)}) = \prod_{k=1}^{N} \frac{\Delta}{\pi} \cdot \frac{1}{1 + \dfrac{(\Delta y^{(k)})^2}{\Delta^2}} \to \max \qquad (1.7.5)$$

Maximizing the likelihood L is equivalent to minimizing its negative logarithm $\psi \stackrel{\text{def}}{=} -\ln(L)$. Differentiating L with respect to Δ and equating the derivative to 0, we get the following formula:

$$\sum_{k=1}^{N} \frac{1}{1 + \dfrac{(\Delta y^{(k)})^2}{\Delta^2}} = \frac{N}{2}. \qquad (1.7.6)$$

To find Δ from this equation, we can use, e.g., the bisection method. Thus, we arrive at the following algorithm.

Monte-Carlo method for estimating the interval uncertainty: algorithm. We select the number of iterations N. For each iteration $k = 1, \ldots, N$, we do the following:

- First, we simulate $\Delta x_i^{(k)}$ based on Cauchy distribution with parameter Δ_i. We can do this, e.g., by computing $\Delta_i^{(k)} = \Delta_i \cdot \tan(\pi \cdot (r_{ik} - 0.5))$, where r_{ik} is the result of a standard random number generator that generates the numbers uniformly distributed on the interval $[0, 1]$.
- After that, we compute the difference

$$\Delta y^{(k)} \stackrel{\text{def}}{=} \widetilde{y} - f(\widetilde{x}_1 - \Delta x_1^{(k)}, \ldots, x_n - \Delta x_n^{(k)}). \qquad (1.7.7)$$

Now, we can find Δ by using bisection to solve the Eq. (1.7.6). Specifically, we start with $\underline{\Delta} = 0$ and $\overline{\Delta} = \max\limits_{1 \le k \le N} |\Delta y^{(k)}|$. For $\Delta = \underline{\Delta}$, the left-hand side of the formula (1.7.6) is smaller than $N/2$, while for $\Delta = \underline{\Delta}$, this left-hand side is larger than $N/2$. Thus, if we want to get Δ with the desired accuracy ε, while $\overline{\Delta} - \underline{\Delta} > \varepsilon$, we do the following:

- we compute $\Delta_{\text{mid}} = \dfrac{\underline{\Delta} + \overline{\Delta}}{2}$;
- we check whether

$$\sum_{k=1}^{N} \frac{1}{1 + \dfrac{\left(\Delta y^{(k)}\right)^2}{\Delta_{\text{mid}}^2}} < \frac{N}{2}; \tag{1.7.8}$$

- if this inequality is true, we replace $\underline{\Delta}$ with the new value Δ_{mid}, leaving $\overline{\Delta}$ unchanged;
- if this inequality is not true, we replace $\overline{\Delta}$ with the new value Δ_{mid}, leaving $\underline{\Delta}$ unchanged.

In both cases, on each iteration, the width of the interval $\left[\underline{\Delta}, \overline{\Delta}\right]$ becomes twice smaller. Thus, in s steps, we decrease this width by a factor of 2^s. So, in a few steps, we get the desired value Δ. For example, to get the width $\le 0.1\%$ of the original one, it is sufficient to perform only 10 iterations of the bisection procedure.

Monte-Carlo method: computation time. In the Monte-Carlo approach, we need $N + 1$ calls to the data processing algorithm f, where N is the number of simulations.

Good news is that, as in statistical methods in general, the needed number of simulation N is determined only by the desired accuracy ε and does not depend on the number of inputs n. For example, to find Δ with relative accuracy 20% and certainty 95% (i.e., in 95% of the cases), it is sufficient to take $n = 200$ [34].

Thus, when the number of inputs n of the data processing algorithm f is large, the Monte-Carlo method for estimating interval uncertainty is much faster than numerical differentiation.

For many practical problems, we can achieve an even faster speed-up. In both methods described in this section, we apply the original algorithm $f(x_1, \ldots, x_n)$ several times: first, to the tuple of nominal values $(\widetilde{x}_1, \ldots, \widetilde{x}_n)$ and then, to several other tuples $(\widetilde{x}_1 + \eta_1, \ldots, \widetilde{x}_n + \eta_n)$ for some small η_i. For example, in the numerical differentiation method, we apply the algorithm f to tuples $(\widetilde{x}_1, \ldots, \widetilde{x}_{i-1}, \widetilde{x}_i + h_i, \widetilde{x}_{i+1}, \ldots, \widetilde{x}_n)$ corresponding to $i = 1, \ldots, n$.

In many practical cases, once we have computed the value $\widetilde{y} = f(\widetilde{x}_1, \ldots, \widetilde{x}_n)$, we can then compute the values

$$f(\widetilde{x}_1 + \eta_1, \ldots, \widetilde{x}_n + \eta_n) \tag{1.7.9}$$

faster than by directly applying the algorithm f to the corresponding tuple. This happens, for example, if the algorithm for computing $f(x_1, \ldots, x_n)$ solves a system of nonlinear equations $F_j(y_1, \ldots, y_k, x_1, \ldots, x_n) = 0$, $1 \le j \le k$, and then returns $y = y_1$.

In this case, once we know the values $\tilde{y}_j$ for which $F_j(\tilde{y}_1, \ldots, \tilde{y}_k, \tilde{x}_1, \ldots, \tilde{x}_n) = 0$, we can find the values $y_j = \tilde{y}_j + \Delta y_j$ for which

$$F_j(\tilde{y}_1 + \Delta y_1, \ldots, \tilde{y}_k + \Delta y_k, \tilde{x}_1 + \eta_1, \ldots, \tilde{x}_n + \eta_n) = 0 \qquad (1.7.10)$$

by linearizing this system into an easy-to-solve system of linear equations

$$\sum_{j'=1}^{k} a_{jj'} \cdot \Delta y_{j'} + \sum_{i=1}^{n} b_{ji} \cdot \eta_i = 0, \qquad (1.7.11)$$

where $a_{jj'} \overset{\text{def}}{=} \dfrac{\partial F_j}{\partial y_{j'}}$ and $b_{ji} \overset{\text{def}}{=} \dfrac{\partial F_j}{\partial x_i}$.

A similar simplification is possible when the value y corresponding to given values $x_1, \ldots, x_n$ comes from solving a system of nonlinear differential equations

$$\frac{dy_j}{dt} = f_j(y_1, \ldots, y_k, x_1, \ldots, x_n). \qquad (1.7.12)$$

In this case, once we find the solution $\tilde{y}_j(t)$ to the system of differential equations

$$\frac{d\tilde{y}_j}{dt} = f_j(\tilde{y}_1, \ldots, \tilde{y}_k, \tilde{x}_1, \ldots, \tilde{x}_n) \qquad (1.7.13)$$

corresponding to the nominal values, we do not need to explicitly solve the modified system of differential equations

$$\frac{dy_j}{dt} = f_j(y_1, \ldots, y_k, \tilde{x}_1 + \eta_1, \ldots, \tilde{x}_n + \eta_n) \qquad (1.7.14)$$

to find the corresponding solution. Instead, we can take into account that the differences η_i are small; thus, the resulting differences $\Delta y_j(t) \overset{\text{def}}{=} y_j(t) - \tilde{y}_j(t)$ are small. So, we can linearize the resulting differential equations

$$\frac{d\Delta y_j(t)}{dt} = f_j(\tilde{y}_1 + \Delta y_1, \ldots, \tilde{y}_k + \Delta y_k, \tilde{x}_1 + \eta_1, \ldots, \tilde{x}_n + \eta_n) -$$

$$f_j(\tilde{y}_1, \ldots, \tilde{y}_k, \tilde{x}_1, \ldots, \tilde{x}_n) \qquad (1.7.15)$$

into easier-to-solve *linear* equations

$$\frac{d\Delta y_j}{dt} = \sum_{j'=1}^{k} a_{jj'} \cdot \Delta y_{j'} + \sum_{i=1}^{n} b_{ji} \cdot \eta_i, \qquad (1.7.16)$$

where $a_{jj'} \stackrel{\text{def}}{=} \dfrac{\partial f_j}{\partial y_{j'}}$ and $b_{ji} \stackrel{\text{def}}{=} \dfrac{\partial f_j}{\partial x_i}$.

This idea – known as *local sensitivity analysis* – is successfully used in many practical applications; see, e.g., [10, 89].

Comment. In some practical situations, some of these deviations are not small. In such situations, we can no longer use linearization, we need to use global optimization techniques of *global sensitivity*; see, e.g., [10, 89].

1.8   Existing Methods for Estimating Fuzzy Uncertainty: Linearization Case

Analysis of the problem. Let us assume that we have fuzzy estimates $\mu_i(\Delta x_i)$ for all the estimation errors Δx_i. In this case, we want to estimate, for every real number Δy, the degree $\mu(\Delta y)$ to which this number is a possible value of data processing inaccuracy.

The value Δy is a possible value of inaccuracy if there exist values Δx_i

- which are possible as inaccuracies of input estimation, and
- for which $= f(x_1, \ldots, x_n)$.

For simplicity, let us use $\min(a, b)$ to describe "and", and $\max(a, b)$ to describe "or". Then, for each combination of values Δx_i, the degree to which all these values are possible is equal to the minimum of the degrees to which each of them is possible:

$$\min(\mu_1(\Delta x_1), \ldots, \mu_n(\Delta x_n)), \tag{1.8.1}$$

and the desired degree Δy is equal to the maximum of these expressions over all possible combinations of Δx_i:

$$\mu(\Delta y) = \max \min(\mu_1(\Delta x_1), \ldots, \mu_n(\Delta x_n)), \tag{1.8.2}$$

where the maximum is taken over all tuples for which $y = f(x_1, \ldots, x_n)$; this expression is known as *Zadeh's extension principle*.

It is know that the computation of this membership function can be simplified if instead of each membership function $\mu(z)$, we consider its α-cuts, i.e., sets $^\alpha z \stackrel{\text{def}}{=} \{z : \mu(z) \geq \alpha\}$ corresponding to different values $\alpha \in [0, 1]$.

It should be mentioned that since $\mu(z)$ is always non-negative, the above definition (with non-strict inequality $\mu(z) \geq \alpha$) does not work for $\alpha = 0$: strictly speaking, all real numbers z satisfy the corresponding inequality. Thus, for $\alpha = 0$, the α-cut is defined slightly differently: as the closure of the set $\{z : \mu(z) > 0\}$ corresponding to strict inequality.

Usually, membership functions correspond to *fuzzy numbers*, i.e., all α-cuts are intervals. Moreover, the α-cuts corresponding to Δx_i are usually symmetric, i.e., have the form $^\alpha \Delta x_i = [-^\alpha \Delta_i, {}^\alpha \Delta_i]$ for appropriate values $^\alpha \Delta_i$.

One can easily check, based on the formula (1.8.2), that $\mu(\Delta) \geq \alpha$ if and only if there exists a tuple $(\Delta x_1, \ldots, \Delta x_n)$ for which

$$\min(\mu_1(\Delta x_1), \ldots, \mu_n(\Delta x_n)) \geq \alpha, \tag{1.8.3}$$

i.e., equivalently, for which $\mu_i(\Delta_i) \geq \alpha$ for all i.

In other words, Δy belongs to the α-cut if and only if it is a possible value of the expression (1.4.5) when Δx_i belongs to the corresponding α-cut $[-^\alpha \Delta_i, {}^\alpha \Delta_i]$.

We already know how to compute the range of such values Δy. Thus, we arrive at the following algorithm for computing the desired α-cut $[-^\alpha \Delta, {}^\alpha \Delta]$.

How to estimate Δy : case of fuzzy uncertainty – resulting formula. In case of fuzzy uncertainty, for every $\alpha \in [0, 1]$, we are given the α-cuts $[-^\alpha \Delta_i, {}^\alpha \Delta_i]$ describing the expert's uncertainty about the estimates $\tilde{x}_i$ and about the model used in data processing.

Based on these α-cuts, we can compute the α-cuts $[-^\alpha \Delta, {}^\alpha \Delta]$ for Δy as follows:

$$^\alpha \Delta = \sum_{i=1}^{n} |c_i| \cdot {}^\alpha \Delta_i. \tag{1.8.4}$$

Comment. In principle, there are infinitely many different values α in the interval $[0, 1]$. However, we should take into account that the values α correspond to experts' degrees of confidence, and experts cannot describe their degrees with too much accuracy.

Usually, it is sufficient to consider only eleven values $\alpha = 0.0$, $\alpha = 0.1$, $\alpha = 0.2, \ldots, \alpha = 0.9$, and $\alpha = 1.0$. Thus, we need to apply the formula (1.8.4) eleven times.

This is in line with the fact that, as psychologists have found, we usually divide each quantity into 7 plus plus minus 2 categories – this is the largest number of categories whose meaning we can immediately grasp; see, e.g., [53, 86]. For some people, this "magical number" is $7 + 2 = 9$, for some it is $7 - 2 = 5$. So, to make sure that do not miss on some people's subtle divisions, it is sufficient to have at least 9 different categories. From this viewpoint, 11 categories are sufficient; usually the above eleven values are chosen since for us, it is easier to understand decimal numbers.

1.9 Open Problems and What We Do in This Book

What we want. In engineering applications, we want methods for estimating uncertainty which are:

- *accurate* – this is most important in most engineering applications;
- *fast* – this is important in some engineering applications where we need real-time computations,

- *understandable* to engineers – otherwise, engineers will be reluctant to use them, and
- sufficiently *general* – so that they can be applied in all kinds of situations.

It is therefore desirable to try to modify the existing methods for estimating uncertainty so that these methods will become more accurate, faster, more understandable, and/or more general.

This book contains the results of our research:

- In Chap. 2, we show how to make these methods more accurate.
- In Chap. 3, we show how to make these methods faster.
- In Chap. 4, we show how to make these methods more understandable to engineers.
- In Chap. 5, we analyze how to make these methods more general.

In Chap. 6, we analyze how to make decisions under uncertainty. The final Chap. 7 contains conclusions.

Chapter 2
How to Get More Accurate Estimates

In most real-life situations, we have all types of uncertainty: probabilistic, interval, fuzzy, and the model inaccuracy. There are methods for processing each of these types of uncertainty.

Often, one type of uncertainty is dominant. In this case, practitioners usually ignore all other types of uncertainty and only take this dominant uncertainty into account. In such situations, it is desirable to analyze how accurate are the resulting estimates – and what can we gain if we use a different type of uncertainty or, better yet, if we take several types of uncertainty into account. In this chapter, we provide several cases of such analysis.

In Sect. 2.1, we show that in many practical situations – especially in problems of system identification – interval estimates can lead to better accuracy than the traditional statistical ones. Of course, an even higher accuracy can be obtained if we take into account both probabilistic and interval uncertainty. This is what we show in Sect. 2.2 – on the example of providing the value which best represents a given sample – and in Sect. 2.3, on the example of taking model inaccuracy into account. Accuracy of fuzzy-based estimates is analyzed in Sect. 2.4.

2.1 In System Identification, Interval Estimates Can Lead to Much Better Accuracy than the Traditional Statistical Ones: General Algorithm and Case Study

In many real-life situations, we know the upper bound of the measurement errors, and we also know that the measurement error is the joint result of several independent small effects. In such cases, due to the Central Limit Theorem, the corresponding probability distribution is close to Gaussian, so it seems reasonable to apply the

© Springer International Publishing AG, part of Springer Nature 2018
A. Pownuk and V. Kreinovich, *Combining Interval, Probabilistic, and Other Types of Uncertainty in Engineering Applications*, Studies in Computational Intelligence 773, https://doi.org/10.1007/978-3-319-91026-0_2

standard Gaussian-based statistical techniques to process this data – in particular, when we need to identify a system. Yes, in doing this, we ignore the information about the bounds, but since the probability of exceeding them is small, we do not expect this to make a big difference on the result. Surprisingly, it turns out that in some practical situations, we get a much more accurate estimates if we, vice versa, take into account the bounds – and ignore all the information about the probabilities. In this section, we explain the corresponding algorithms. and we show, on a practical example, that using this algorithm can indeed lead to a drastic improvement in estimation accuracy.

The results of this section first appeared in [41].

2.1.1 Formulation of the Problem

System identification: a general problem. In many practical situations, we are interested in a quantity y which is difficult – or even impossible – to measure directly. This difficulty and/or impossibility may be technical: e.g.:

- while we can directly measure the distance between the two buildings by simply walking there,
- there is no easy way to measure the distance to a nearby start by flying there.

In other cases, the impossibility comes from the fact that we are interested in predictions – and, of course, today we cannot measure tomorrow's temperature.

To estimate the value of such a difficult-to-directly-measure quantity y, a natural idea is:

- to find easier-to-measure quantities $x_1, \ldots, x_n$ that are related to y by a known dependence $y = f(x_1, \ldots, x_n)$, and then
- to use the results $\widetilde{x}_i$ of measuring these auxiliary quantities to estimate y as

$$\widetilde{y} \overset{\text{def}}{=} f(\widetilde{x}_1, \ldots, \widetilde{x}_n).$$

For example:

- We can find the distance to a nearby star by measuring the direction to this star in two seasons, when the Earth is at different sides of the Sun, and the angle is thus slightly different.
- To predict tomorrow's temperature, we can measure the temperature and wind speed and direction at different locations today, and use the general equations for atmospheric dynamics to estimate tomorrow's temperature.

In some cases, we already know the dependence $y = f(x_1, \ldots, x_n)$. In many other situations, we know the general form of this dependence, but there are some parameters that we need to determine experimentally. In other words, we know that

$$y = f(a_1, \ldots, a_m, x_1, \ldots, x_n) \tag{2.1.1}$$

for some parameters $a_1, \ldots, a_m$ that need to be experimentally determined.

For example, we may know that the dependence of y on x_1 is linear, i.e., $y = a \cdot x_1 + b$, but we do not know the exact values of the corresponding parameters a and b.

In general, the problem of finding the parameters a_j is known as the problem of *system identification*.

What information we use for system identification. To identify a system, i.e., to find the values of the parameters a_j, we can use the results $\widetilde{y}_k$ and $\widetilde{x}_{ki}$ of measuring the quantities y and x_i in several different situations $k = 1, \ldots, K$.

How do we identify the system: need to take measurement uncertainty into account. Most information comes from measurements, but measurements are not 100% accurate: in general, the measurement result $\widetilde{x}$ is somewhat different from the actual (unknown) value x of the corresponding quantity: $\Delta x \overset{\text{def}}{=} \widetilde{x} - x \neq 0$; see, e.g., [82].

As a result, while we know that for every k, the corresponding (unknown) exact values y_k and x_{ki} are related by the dependence (2.1.1):

$$y_k = f(a_1, \ldots, a_m, x_{k1}, \ldots, x_{kn}), \tag{2.1.2}$$

a similar relation between the approximate values $\widetilde{y}_k \approx y_k$ and $\widetilde{x}_{ki} \approx x_{ki}$ is only approximate:

$$\widetilde{y}_k \approx f(a_1, \ldots, a_m, \widetilde{x}_{k1}, \ldots, \widetilde{x}_{kn}).$$

It is therefore important to take this uncertainty into account when estimating the values of the parameters $a_1, \ldots, a_m$.

How can we describe the uncertainty? In all the cases, we should know the bound Δ on the absolute value of the measurement error: $|\Delta x| \leq \Delta$; see, e.g., [82]. This means that only values Δx from the interval $[-\Delta, \Delta]$ are possible.

If this is the only information we have then, based on the measurement result $\widetilde{x}$, the only information that we have about the unknown actual value x is that this value belongs to the interval $[\widetilde{x} - \Delta, \widetilde{x} + \Delta]$. There are many techniques for processing data under such interval uncertainty; this is known as *interval computations*; see, e.g., [26, 58].

Ideally, it is also desirable to know how frequent are different values Δx within this interval. In other words, it is desirable to know the probabilities of different values $\Delta x \in [-\Delta, \Delta]$.

A usual way to get these probabilities is to take into account that in many measurement situations, the measurement uncertainty Δx comes from many different independent sources. It is known that for large N, the distribution of the sum of N independent random variables becomes close to the normal (Gaussian) distribution – and tends to it when $N \to \infty$. This result – known as the Central Limit Theorem (see, e.g., [97]) – explain the ubiquity of normal distributions. It is therefore reasonable to assume that the actual distribution is Gaussian – and this is what most practitioners do in such situations [82].

Two approximations, two options. A seemingly minor problem with the Gaussian distribution is that it is, strictly speaking, *not* located on any interval: for this distribution, the probability of measurement error Δx to be in any interval – no matter how far away from Δ – is non-zero.

From this viewpoint, the assumption that the distribution is Gaussian is an approximation. It seems like a very good approximation, since for normal distribution with means 0 and standard deviation σ:

- the probability to be outside the 3σ interval $[-3\sigma, 3\sigma]$ is very small, approximately 0.1%, and
- the probability for it to be outside the 6σ interval is about 10^{-8}, practically negligible.

Yes, when we use Gaussian distributions, we ignore the information about the bounds, but, at first glance, since the difference is small, this should not affect the measurement results.

At first glance, the opposite case – when we keep the bounds but ignore all the information about probabilities, maybe add imprecise (fuzzy) expert information about possible values of Δx (see, e.g., [28, 61, 108]) – should be much worse.

What we found. Our results show, somewhat surprisingly, that the opposite is true: that if ignore the probabilistic information and use only interval (or fuzzy) information, we get much more accurate estimates for the parameters a_j than in the usual statistical methodology.

This may not be fully surprising, since there are theoretical results showing that asymptotically, interval bounds can be better; see, e.g., [104]. However, the drastic improvement in accuracy was somewhat unexpected.

The structure of the section. First, we describe the algorithm that we used, both the general algorithm and the specific algorithm corresponding to the linear case. After that, we show the results of applying this algorithm.

2.1.2 System Identification Under Interval Uncertainty: General Algorithm

Formulation of the problem in the interval case. For each pattern $k = 1, \ldots, K$, we know the measurement results $\widetilde{y}_k$ and $\widetilde{x}_{ki}$, and we know the accuracies Δ_k and Δ_{ki} of the corresponding measurements. Thus, we know that:

- the actual (unknown) value y_k belongs to the interval

$$[\underline{y}_k, \overline{y}_k] = [\widetilde{y}_k - \Delta_k, \widetilde{y}_k + \Delta_k];$$

and
- the actual (unknown) value x_{ki} belongs to the interval

$$[\underline{x}_{ki}, \overline{x}_{ki}] = [\widetilde{x}_{ki} - \Delta_{ki}, \widetilde{x}_{ki} + \Delta_{ki}].$$

We need to find the values $a_1, \ldots, a_m$ for which, for every k, for some values $x_{ki} \in [\underline{x}_{ki}, \overline{x}_{ki}]$, the quantity $f(a_1, \ldots, a_m, x_{k1}, \ldots, x_{kn})$ belongs to the interval $[\underline{y}_k, \overline{y}_k]$.

Specifically, for each j from 1 to m, we would like to find the range $[\underline{a}_j, \overline{a}_j]$ of all possible values of the corresponding parameter a_j.

What happens in the statistical case. In the statistical case, we use the Least Squares method [97] and find the values $\widetilde{a}_1, \ldots, \widetilde{a}_m$ that minimize the sum of the squares of all the discrepancies:

$$\sum_{k=1}^{K} (\widetilde{y}_k - f(a_1, \ldots, a_m, \widetilde{x}_{k1}, \ldots, \widetilde{x}_{kn}))^2 \to \min_{a_1, \ldots, a_m}.$$

Possibility of linearization. Let us denote $\Delta a_j \stackrel{\text{def}}{=} \widetilde{a}_j - a_j$, where $\widetilde{a}_j$ are the least-squares estimates. In these terms, we have $a_j = \widetilde{a}_j - \Delta a_j$ and $x_{ki} = \widetilde{x}_{ki} - \Delta x_{ki}$. Thus, the corresponding value y_k has the form

$$y_k = f(a_1, \ldots, a_n, x_{k1}, \ldots, x_{kn}) = \tag{2.1.3}$$

$$f(\widetilde{a}_1 - \Delta a_1, \ldots, \widetilde{a}_m - \Delta a_m, \widetilde{x}_{k1} - \Delta x_{k1}, \ldots, \widetilde{x}_{kn} - \Delta x_{kn}).$$

The measurement errors Δx_{ki} are usually relatively small. As a result, the difference between the least-squared values $\widetilde{a}_j$ and the actual (unknown) values a_j is also small. Thus, we can expand the expression (2.1.3) in Taylor series and keep only linear terms in this expansion. This results in:

$$y_k = Y_k - \sum_{j=1}^{m} b_j \cdot \Delta a_j - \sum_{i=1}^{n} b_{ki} \cdot \Delta x_{ki}, \tag{2.1.4}$$

where we denoted

$$Y_k \stackrel{\text{def}}{=} f(\widetilde{a}_1, \ldots, \widetilde{a}_m, \widetilde{x}_{k1}, \ldots, \widetilde{x}_{kn}), \tag{2.1.5}$$

$$b_j \stackrel{\text{def}}{=} \frac{\partial f}{\partial a_j}_{|a_1 = \widetilde{a}_1, \ldots, a_m = \widetilde{a}_m, x_{k1} = \widetilde{x}_{k1}, \ldots, x_{km} = \widetilde{x}_{kn}}, \tag{2.1.6}$$

and

$$b_{ki} \overset{\text{def}}{=} \frac{\partial f}{\partial x_{ki}}_{|a_1=\tilde{a}_1,\dots,a_m=\tilde{a}_m,x_{k1}=\tilde{x}_{k1},\dots,x_{kn}=\tilde{x}_{kn}} . \tag{2.1.7}$$

We want to make sure that for some $\Delta x_{ki} \in [-\Delta_{ki}, \Delta_{ki}]$, the value (2.1.4) belongs to the interval $[\underline{y}_k, \overline{y}_k]$. Thus, we want to make sure that for each k, the range $[\underline{Y}_k, \overline{Y}_k]$ of all possible values of the expression (2.1.4) when $\Delta x_{ki} \in [-\Delta_{ki}, \Delta_{ki}]$ has a non-empty intersection with the interval $[\underline{y}_k, \overline{y}_k]$.

Let us thus find the expression for the range $[\underline{Y}_k, \overline{Y}_k]$. One can easily see that when $\Delta x_{ki} \in [-\Delta_{ki}, \Delta_{ki}]$, the value of the product $b_{ki} \cdot \Delta x_{ki}$ ranges from $-|b_{ki}| \cdot \Delta_{ki}$ to $|b_{ki}| \cdot \Delta_{ki}$. Thus, the smallest possible value $\underline{Y}_k$ and the largest possible value $\overline{Y}_k$ of the expression (2.1.4) are equal to:

$$\underline{Y}_k = Y_k - \sum_{j=1}^{m} b_j \cdot \Delta a_j - \sum_{i=1}^{n} |b_{ki}| \cdot \Delta_{ki}, \tag{2.1.8}$$

and

$$\overline{Y}_k = Y_k - \sum_{j=1}^{m} b_j \cdot \Delta a_j + \sum_{i=1}^{n} |b_{ki}| \cdot \Delta_{ki}. \tag{2.1.9}$$

One can easily check that the two intervals $[\underline{y}_k, \overline{y}_k]$ and $[\underline{Y}_k, \overline{Y}_k]$ intersect if and only if:

- the lower endpoint of the first interval does not exceed the upper endpoint of the second interval, and
- the lower endpoint of the second interval does not exceed the upper endpoint of the first interval,

i.e., if $\underline{y}_k \le \overline{Y}_k$ and $\underline{Y}_k \le \overline{y}_k$.

These equalities are linear in terms of the unknowns. So, the corresponding problem of finding the smallest and largest possible values of a_j becomes a particular case of optimizing a linear function under linear inequalities. For this class of problems – known as *linear programming* problems – there are known efficient algorithms; see, e.g., [48].

Thus, we arrive at the following algorithm.

Resulting algorithm. We are given:

- the expression $f(a_1, \dots, a_m, x_1, \dots, x_n)$ with unknown parameters a_j, and
- K measurement patterns.

For each pattern k, we know:

- the measurement results $\tilde{y}_k$ and $\tilde{x}_{ki}$, and
- the accuracies Δ_k and Δ_{ki} of the corresponding measurements.

Based on these inputs, we first use the Least Squares method to find the estimates $\tilde{a}_1, \ldots, \tilde{a}_m$. Then, we compute the values $\underline{y}_k = \tilde{y}_k - \Delta_k$, $\overline{y}_k = \tilde{y}_k + \Delta_k$, and the values (2.1.5)–(2.1.7).

After that, for each j_0, we find the desired value $\underline{a}_{j_0}$ as the solution to the following linear programming problem: minimize a_{j_0} under the constraints that for all k, we have

$$\underline{y}_k \leq Y_k - \sum_{j=1}^{m} b_j \cdot \Delta a_j + \sum_{i=1}^{n} |b_{ki}| \cdot \Delta_{ki}$$

and

$$Y_k - \sum_{j=1}^{m} b_j \cdot \Delta a_j - \sum_{i=1}^{n} |b_{ki}| \cdot \Delta_{ki} \leq \overline{y}_k.$$

The value $\overline{a}_{j_0}$ can be found if we maximize a_{j_0} under the same $2K$ constraints.

How to use these formulas to estimate y? What if we now need to predict the value y corresponding to given values $x_1, \ldots, x_m$? In this case,

$$y = f(a_1, \ldots, a_m, x_1, \ldots, x_n) =$$

$$f(\tilde{a}_1 - \Delta a_1, \ldots, \tilde{a}_m - \Delta a_m, x_1, \ldots, x_n) =$$

$$\tilde{y} - \sum_{j=1}^{M} B_j \cdot \Delta a_j,$$

where we denoted

$$\tilde{y} = f(\tilde{a}_1, \ldots, \tilde{a}_m, x_1, \ldots, x_n)$$

and

$$B_j \overset{\text{def}}{=} \frac{\partial f}{\partial a_j}\bigg|_{a_1 = \tilde{a}_1, \ldots, a_m = \tilde{a}_m, x_1, \ldots, x_n}.$$

In this case:

- the smallest possible value $\underline{y}$ of y can be found by minimizing the linear combination $\tilde{y} - \sum_{j=1}^{m} B_j \cdot \Delta a_j$ under the above constraints; and
- the largest possible value $\overline{y}$ of y can be found by maximizing the same linear combination $\tilde{y} - \sum_{j=1}^{m} B_j \cdot \Delta a_j$ under the above constraints.

What if we underestimated the measurement inaccuracy? When we applied this algorithm to several specific situations, in some cases, to our surprise, it turned out that the constraints were inconsistent. This means that we underestimated the measurement inaccuracy.

Since measuring y is the most difficult part, most probably we underestimated the accuracies of measuring y. If we denote the ignored part of the y-measuring error by ε, this means that, instead of the original bounds Δ_k on $|\Delta y_k|$, we should have bounds $\Delta_k + \varepsilon$. In this case:

- instead of the original values $\underline{y}_k = \widetilde{y}_k - \Delta_k$ and $\overline{y}_k = \widetilde{y}_k + \Delta_k$,
- we should have new bounds $\widetilde{y}_k - \Delta_k - \varepsilon$ and $\widetilde{y}_k + \Delta_k + \varepsilon$.

It is reasonable to look for the smallest possible values $\varepsilon > 0$ for which the constrains will become consistent. Thus, we arrive at the following linear programming problem: minimize $\varepsilon > 0$ under the constraints

$$\widetilde{y}_k - \Delta_k - \varepsilon \leq Y_k - \sum_{j=1}^{m} b_j \cdot \Delta a_j + \sum_{i=1}^{n} |b_{ki}| \cdot \Delta_{ki}$$

and

$$Y_k - \sum_{j=1}^{m} b_j \cdot \Delta a_j - \sum_{i=1}^{n} |b_{ki}| \cdot \Delta_{ki} \leq \widetilde{y}_k + \Delta_k + \varepsilon.$$

2.1.3 System Identification Under Interval Uncertainty: Simplest Case of Linear Dependence on One Variable

Description of the simplest case. Let us consider the simplest case when there is only one variable x (i.e., $n = 1$), and the dependence on this variable is linear, i.e.,

$$y = a \cdot x + b.$$

In this case:

- we have K measurement results $\widetilde{x}_k$ with accuracy Δ_k, resulting in intervals $[\underline{x}_k, \overline{x}_k]$, and similarly,
- we have intervals $[\underline{y}_k, \overline{y}_k]$ of possible values of y_k.

Based on this information, we need to find the set of all possible values of the pairs (a, b). In particular, we need to find the ranges of possible values of a and b.

Why we need to consider this case separately. Linear programming is feasible, but its algorithms are intended for a general case and thus, for the case when we have few unknowns, usually run for too long. In such situations, it is often possible to find faster techniques.

Finding bounds on a: analysis of the problem. Let us first consider the cases when $a > 0$. (The case when $a < 0$ can be handled similarly – or the same by replacing a with $-a$ and x_k with $-x_k$.)

In this case, the set of possible values of $a \cdot x_k + b$ when $x_k \in [\underline{x}_k, \overline{x}_k]$ has the form

$$[a \cdot \underline{x}_k + b, a \cdot \overline{x}_k + b].$$

We want to make sure that this interval intersects with $[\underline{y}_k, \overline{y}_k]$, i.e., that for every k, we have

$$a \cdot \underline{x}_k + b \leq \overline{y}_k \text{ and } \underline{y}_k \leq a \cdot \overline{x}_k + b.$$

Thus, once we know a, we have the following lower bounds and upper bounds for b:

$$\underline{y}_k - a \cdot \overline{x}_k \leq b \text{ and } b \leq \overline{y}_k - a \cdot \underline{x}_k.$$

Such a value b exists if and only if every lower bound for b is smaller than or equal to every upper bound for b, i.e., if and only if, for every k and ℓ, we have

$$\underline{y}_k - a \cdot \overline{x}_k \leq \overline{y}_\ell - a \cdot \underline{x}_\ell,$$

i.e., equivalently,

$$\overline{y}_\ell - \underline{y}_k \geq a \cdot (\underline{x}_\ell - \overline{x}_k).$$

- When the difference $\underline{x}_\ell - \overline{x}_k$ is positive, we divide the above inequality by this difference and get an upper bound on a:

$$a \leq \frac{\overline{y}_\ell - \underline{y}_k}{\underline{x}_\ell - \overline{x}_k};$$

- When this difference is negative, after division, we get a lower bound on a:

$$a \geq \frac{\overline{y}_\ell - \underline{y}_k}{\underline{x}_\ell - \overline{x}_k}.$$

Thus, the range $[\underline{a}, \overline{a}]$ for a goes from the largest of the lower bounds to the smallest of the upper bounds. So, we arrive at the following formulas.

Resulting range for a. The resulting range for a is $[\underline{a}, \overline{a}]$, where:

$$\underline{a} = \max_{k,\ell:\ \underline{x}_\ell < \overline{x}_k} \frac{\overline{y}_\ell - \underline{y}_k}{\underline{x}_\ell - \overline{x}_k};$$

$$\overline{a} = \min_{k,\ell:\ \underline{x}_\ell > \overline{x}_k} \frac{\overline{y}_\ell - \underline{y}_k}{\underline{x}_\ell - \overline{x}_k}.$$

Range for b: analysis of the problem. For $a > 0$, we need to satisfy, for each k, the inequalities

$$a \cdot \underline{x}_k + b \leq \overline{y}_k \text{ and } \underline{y}_k \leq a \cdot \overline{x}_k + b.$$

Equivalently, we get

$$a \cdot x_k \leq \overline{y}_k - b \text{ and } \overline{y}_k - b \leq a \cdot \overline{x}_k.$$

By dividing these inequalities by a coefficient at a, we have the following bounds for a:

- for all k for which $\underline{x}_k > 0$, we get an upper bound

$$a \leq \frac{\overline{y}_k}{\underline{x}_k} - \frac{1}{\underline{x}_k} \cdot b;$$

- for all k for which $\underline{x}_k < 0$, we get a lower bound

$$\frac{\overline{y}_k}{\underline{x}_k} - \frac{1}{\underline{x}_k} \cdot b \leq a;$$

- for all k for which $\overline{x}_k > 0$, we get a lower bound

$$\frac{\underline{y}_k}{\overline{x}_k} - \frac{1}{\overline{x}_k} \cdot b \leq a;$$

- for all k for which $\overline{x}_k > 0$, we get an upper bound

$$a \leq \frac{\underline{y}_k}{\overline{x}_k} - \frac{1}{\overline{x}_k} \cdot b.$$

Thus, we get lower bounds $A_p + B_p \cdot b \leq a$ and upper bounds $a \leq C_q + D_q \cdot b$. These inequalities are consistent if every lower bound is smaller than or equal than every upper bound, i.e., when $A_p + B_p \cdot b \leq C_q + D_q \cdot b$, or, equivalently, when

$$(D_q - B_p) \cdot b \geq A_p - C_q.$$

So, similarly to the a-case, we arrive at the following formulas:

Resulting range for b. The range for b is equal to $[\underline{b}, \overline{b}]$, where

$$\underline{b} = \max_{p,q: \ D_q > B_p} \frac{A_p - C_q}{D_q - B_p};$$

$$\overline{b} = \max_{p,q: \ D_q < B_p} \frac{A_p - C_q}{D_q - B_p}.$$

What if we underestimated the measurement inaccuracy. In this case, instead of the original bounds $\underline{y}_k$ and $\overline{y}_k$, we get the new bounds $\underline{y}_k - \varepsilon$ and $\overline{y}_k + \varepsilon$. Thus, instead of the original difference $\overline{y}_\ell - \underline{x}_k$, we get a new difference $(\overline{y}_\ell - \underline{y}_k) - \varepsilon$. The lower and upper bounds for a are thus as follows:

- When the difference $\underline{x}_\ell > \overline{x}_k$, we get

$$a \leq \frac{\overline{y}_\ell - \underline{y}_k}{\underline{x}_\ell - \overline{x}_k} + \frac{2}{\underline{x}_\ell - \overline{x}_k} \cdot \varepsilon;$$

- When this difference is negative, after division, we get a lower bound on a:

$$a \geq \frac{\overline{y}_\ell - \underline{y}_k}{\underline{x}_\ell - \overline{x}_k} + \frac{2}{\underline{x}_\ell - \overline{x}_k} \cdot \varepsilon; .$$

Thus, we get lower bounds $A_p + B_p \cdot \varepsilon \leq a$ and upper bounds $a \leq C_q + D_q \cdot \varepsilon$. These inequalities are consistent if every lower bound is smaller than or equal than every upper bound, i.e., when $A_p + B_p \cdot \varepsilon \leq C_q + D_q \cdot \varepsilon$, or, equivalently, when $(D_q - B_p) \cdot \varepsilon \geq A_p - C_q$. So, similarly to the a- an b-cases, we arrive at the following formulas.

The desired lower bound for ε for b is equal to the largest of the lower bounds, i.e., to

$$\varepsilon = \max_{p,q:\ D_q > B_p} \frac{A_p - C_q}{D_q - B_p}.$$

2.1.4   Case Study

Description of the case study. One of the important engineering problems is the problem of storing energy. For example, solar power and wind turbines provide access to large amounts of renewable energy, but this energy is not always available – the sun goes down, the wind dies – and storing it is difficult. Similarly, electric cars are clean, but the need to store energy forces us to spend a lot of weight on the batteries.

Therefore, it is desirable to develop batteries with high energy density. One of the most promising directions is using molten salt batteries, including liquid metal batteries. These batteries offer high energy density and high power density.

To properly design these batteries, we need to analyze how the heat of fusion – i.e., the energy needed to melt the material – depends on the melting temperature. It is known that this dependence is linear.

Results. On Fig. 2.1, we show the results of our analysis.

It turns out that the bounds on y coming from our method are an order of magnitude smaller that the 2σ-bounds coming from the traditional statistical analysis; see [85]

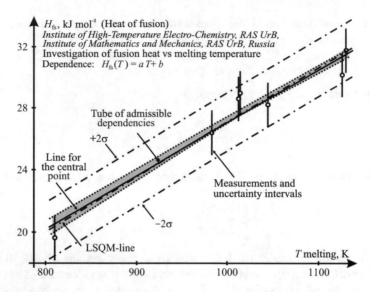

Fig. 2.1 Case when interval uncertainty leads to much more accurate estimates than the probabilistic one

for details. The paper [85] also contains the description of the set of all possible pairs (a, b), i.e., all pairs which are consistent with all the measurement results.

A similar improvement was observed in other applications as well. A similar – albeit not so drastic – improvement was observed in other applications ranging from catalysis and to mechanics; see, e.g., [3, 22, 40, 41, 43, 68].

2.1.5 Conclusions

Traditional engineering techniques for estimating uncertainty of the results of data processing are based on the assumption that the measurement errors are normally distributed. In practice, the distribution of measurement errors is indeed often close to normal, so, in principle, we can use the traditional techniques to gauge the corresponding uncertainty.

In many practical situations, however, we also have an additional information about measurement uncertainty: namely, we also know the upper bounds Δ on the corresponding measurement errors. As a result, once we know the measurement result $\tilde{x}$, we can compute the interval $[\tilde{x} - \Delta, \tilde{x} + \Delta]$ which is guaranteed to contain the actual (unknown) value of the measured quantity. Once we know these intervals, we can use interval computations techniques to estimate the accuracy of the result of data processing. For example, for linear models, we can use linear programming techniques to compute the corresponding bounds.

Which of the two approaches lead to more accurate estimate:

- the traditional approach, in which we ignore the upper bounds and only consider the probability distributions, or
- the interval approach, in which we only take into account the upper bounds and ignore the probabilistic information?

Previous theoretical analysis (see, e.g., [104]) shows that, in general, asymptotically, when the number of measurements n increases, the interval estimates become more accurate than the probabilistic ones.

In this section, we show, on the example of system identification, that for several reasonable practical situations, interval techniques indeed lead to much more accurate estimates than the statistical ones – even when we have only 7 measurement results. Thus, our recommendation is that in situations when we also know upper bounds, and we have a reasonable number of measurement results, it is beneficial to use interval techniques – since they lead to more accurate estimates.

For linear models with two parameters, we also provide a new interval-based algorithm for finding the ranges of these parameters, an algorithm which is much faster than the general linear programming techniques.

2.2   Which Value $\widetilde{x}$ Best Represents a Sample $x_1, \ldots, x_n$: Utility-Based Approach Under Interval Uncertainty

In many practical situations, we have several estimates $x_1, \ldots, x_n$ of the same quantity x. In such situations, it is desirable to combine this information into a single estimate $\widetilde{x}$. Often, the estimates x_i come with interval uncertainty, i.e., instead of the exact values x_i, we only know the intervals $[\underline{x}_i, \overline{x}_i]$ containing these values. In this section, we formalize the problem of finding the combined estimate $\widetilde{x}$ as the problem of maximizing the corresponding utility, and we provide an efficient (quadratic-time) algorithm for computing the resulting estimate.

Results described in this section first appeared in [79].

2.2.1   Which Value $\widetilde{x}$ Best Represents a Sample $x_1, \ldots, x_n$: Case of Exact Estimates

Need to combine several estimates. In many practical situations, we have several estimates $x_1, \ldots, x_n$ of the same quantity x. In such situations, it is often desirable to combine this information into a single estimate $\widetilde{x}$; see, e.g., [82].

Probabilistic case. If we know the probability distribution of the corresponding estimation errors $x_i - x$, then we can use known statistical techniques to find $\widetilde{x}$, e.g., we can use the Maximum Likelihood Method; see, e.g., [97].

Need to go beyond the probabilistic case. In many cases, however, we do not have any information about the corresponding probability distribution [82]. How can we then find $\tilde{x}$?

Utility-based approach. According to the general decision theory, decisions of a rational person are equivalent to maximizing his/her *utility value u*; see, e.g., [19, 46, 59, 84]. Let us thus find the estimate $\tilde{x}$ for which the utility $u(\tilde{x})$ is the largest.

Our objective is to use a single value $\tilde{x}$ instead of all n values x_i. For each i, the disutility $d = -u$ comes from the fact that if the actual estimate is x_i and we use a different value $\tilde{x} \neq x_i$ instead, we are not doing an optimal thing. For example, if the optimal speed at which the car needs the least amount of fuel is x_i, and we instead run it at a speed $\tilde{x} \neq x_i$, we thus waste some fuel.

For each i, the disutility d comes from the fact that the difference $\tilde{x} - x_i$ is different from 0; there is no disutility if we use the actual value, so $d = d(\tilde{x} - x_i)$ for an appropriate function $d(y)$, where $d(0) = 0$ and $d(y) > 0$ for $y \neq 0$.

The estimates are usually reasonably accurate, so the difference $x_i - \tilde{x}$ is small, and we can expand the function $d(y)$ in Taylor series and keep only the first few terms in this expansion:

$$d(y) = d_0 + d_1 \cdot y + d_2 \cdot y^2 + \cdots$$

From $d(0) = 0$ we conclude that $d_0 = 0$. From $d(y) > 0$ for $y \neq 0$ we conclude that $d_1 = 0$ (else we would have $d(y) < 0$ for some small y) and $d_2 > 0$, so $d(y) = d_2 \cdot y^2 = d_2 \cdot (\tilde{x} - x_i)^2$.

The overall disutility $d(\tilde{x})$ of using $\tilde{x}$ instead of each of the values $x_1, \ldots, x_n$ can be computed as the sum of the corresponding disutilities

$$d(\tilde{x}) = \sum_{i=1}^{n} d(\tilde{x} - x_i)^2 = d_2 \cdot \sum_{i=1}^{n} (\tilde{x} - x_i)^2.$$

Maximizing utility $u(\tilde{x}) \overset{\text{def}}{=} -d(\tilde{x})$ is equivalent to minimizing disutility.

The resulting combined value. Since $d_2 > 0$, minimizing the disutility function is equivalent to minimizing the re-scaled disutility function

$$D(\tilde{x}) \overset{\text{def}}{=} \frac{d(\tilde{x})}{d_2} = \sum_{i=1}^{n} (\tilde{x} - x_i)^2.$$

Differentiating this expression with respect to $\tilde{x}$ and equating the derivative to 0, we get

$$\tilde{x} = \frac{1}{n} \cdot \sum_{i=1}^{n} x_i.$$

This is the well-known sample mean.

2.2.2 Case of Interval Uncertainty: Formulation of the Problem

Formulation of the practical problem. In many practical situations, instead of the exact estimates x_i, we only know the intervals $[\underline{x}_i, \overline{x}_i]$ that contain the unknown values x_i. How do we select the value x in this case?

Towards precise formulation of the problem. For different values x_i from the corresponding intervals $[\underline{x}_i, \overline{x}_i]$, we get, in general, different values of utility

$$U(\tilde{x}, x_1, \ldots, x_n) = -D(\tilde{x}, x_1, \ldots, x_n),$$

where $D(\tilde{x}, x_1, \ldots, x_n) = \sum_{i=1}^{n} (\tilde{x} - x_i)^2$. Thus, all we know is that the actual (unknown) value of the utility belongs to the interval $[\underline{U}(\tilde{x}), \overline{U}(\tilde{x})] = [-\overline{D}(\tilde{x}), -\underline{D}(\tilde{x})]$, where

$$\underline{D}(\tilde{x}) = \min D(\tilde{x}, x_1, \ldots, x_n),$$

$$\overline{D}(\tilde{x}) = \max D(\tilde{x}, x_1, \ldots, x_n),$$

and min and max are taken over all possible combinations of values $x_i \in [\underline{x}_i, \overline{x}_i]$.

In such situations of interval uncertainty, decision making theory recommends using Hurwicz optimism-pessimism criterion [25, 32, 46], i.e., maximize the value

$$U(\tilde{x}) \stackrel{\text{def}}{=} \alpha \cdot \overline{U}(\tilde{x}) + (1 - \alpha) \cdot \underline{U}(\tilde{x}),$$

where the parameter $\alpha \in [0, 1]$ describes the decision maker's degree of optimism. For $U = -D$, this is equivalent to minimizing the expression

$$D(\tilde{x}) = -U(\tilde{x}) = \alpha \cdot \underline{D}(\tilde{x}) + (1 - \alpha) \cdot \overline{D}(\tilde{x}).$$

What we do in this section. In this section, we describe an efficient algorithm for computing the value $\tilde{x}$ that minimizes the resulting objective function $D(\tilde{x})$.

2.2.3 Analysis of the Problem

Let us simplify the expressions for $\underline{D}(\tilde{x})$, $\overline{D}(\tilde{x})$, and $D(\tilde{x})$. Each term $(\tilde{x} - x_i)^2$ in the sum $D(\tilde{x}, x_1, \ldots, x_n)$ depends only on its own variable x_i. Thus, with respect to x_i:

- the sum is the smallest when each of these terms is the smallest, and
- the sum is the largest when each term is the largest.

One can easily see that when x_i is in the $[\underline{x}_i, \overline{x}_i]$, the maximum of a term $(\tilde{x} - x_i)^2$ is always attained at one of the interval's endpoints:

- at $x_i = \underline{x}_i$ when $\tilde{x} \geq \tilde{x}_i \overset{\text{def}}{=} \dfrac{\underline{x}_i + \overline{x}_i}{2}$ and
- at $x_i = \overline{x}_i$ when $\tilde{x} < \tilde{x}_i$.

Thus,

$$\overline{D}(\tilde{x}) = \sum_{i:\tilde{x} < \tilde{x}_i} (\tilde{x} - \overline{x}_i)^2 + \sum_{i:\tilde{x} \geq \tilde{x}_i} (\tilde{x} - \underline{x}_i)^2.$$

Similarly, the minimum of the term $(\tilde{x} - x_i)^2$ is attained:

- for $x_i = \tilde{x}$ when $\tilde{x} \in [\underline{x}_i, \overline{x}_i]$ (in this case, the minimum is 0);
- for $x_i = \underline{x}_i$ when $\tilde{x} < \underline{x}_i$; and
- for $x_i = \overline{x}_i$ when $\tilde{x} > \overline{x}_i$.

Thus,

$$\underline{D}(\tilde{x}) = \sum_{i:\tilde{x} > \overline{x}_i} (\tilde{x} - \overline{x}_i)^2 + \sum_{i:\tilde{x} < \underline{x}_i} (\tilde{x} - \underline{x}_i)^2.$$

So, for $D(\tilde{x}) = \alpha \cdot \underline{D}(\tilde{x}) + (1 - \alpha) \cdot \overline{D}(\tilde{x})$, we get

$$D(\tilde{x}) = \alpha \cdot \sum_{i:\tilde{x} > \overline{x}_i} (\tilde{x} - \overline{x}_i)^2 + \alpha \cdot \sum_{i:\tilde{x} < \underline{x}_i} (\tilde{x} - \underline{x}_i)^2 +$$

$$(1 - \alpha) \cdot \sum_{i:\tilde{x} < \tilde{x}_i} (\tilde{x} - \overline{x}_i)^2 + (1 - \alpha) \cdot \sum_{i:\tilde{x} \geq \tilde{x}_i} (\tilde{x} - \underline{x}_i)^2. \tag{2.2.1}$$

Towards an algorithm. The presence or absence of different values in the above expression depends on the relation of $\tilde{x}$ with respect to the values $\underline{x}_i, \overline{x}_i$, and $\tilde{x}_i$. Thus, if we sort these $3n$ values into a sequence $s_1 \leq s_2 \leq \cdots \leq s_{3n}$, then on each interval $[s_j, s_{j+1}]$, the function $D(\tilde{x})$ is simply a quadratic function of $\tilde{x}$.

A quadratic function attains its minimum on an interval either at one of its midpoints, or at a point when the derivative is equal to 0 (if this point is inside the given interval). Differentiating the above expression for $D(\tilde{x})$, equating the derivative to 0, dividing both sides by 0, and moving terms proportional not containing $\tilde{x}$ to the right-hand side, we conclude that

$$(\alpha \cdot \#\{i : \tilde{x} < \underline{x}_i \text{ or } \tilde{x} > \overline{x}_i\} + 1 - \alpha) \cdot \tilde{x} =$$

$$\alpha \cdot \sum_{i:\tilde{x} > \overline{x}_i} \overline{x}_i + \alpha \cdot \sum_{i:\tilde{x} < \underline{x}_i} \underline{x}_i + (1 - \alpha) \cdot \sum_{i:\tilde{x} < \tilde{x}_i} \overline{x}_i + (1 - \alpha) \cdot \sum_{i:\tilde{x} \geq \tilde{x}_i} \underline{x}_i.$$

Since s_j is a listing of all thresholds values $\underline{x}_i$, $\overline{x}_i$, and $\widetilde{x}_i$, then for $\widetilde{x} \in (s_j, s_{j+1})$, the inequality $\widetilde{x} < \underline{x}_i$ is equivalent to $s_{j+1} \leq \underline{x}_i$. Similarly, the inequality $\widetilde{x} > \underline{x}_i$ is equivalent to $s_j \geq \overline{x}_i$. In general, for values $\widetilde{x} \in (s_j, s_{j+1})$, the above equation gets the form

$$(\alpha \cdot \#\{i : \widetilde{x} < \underline{x}_i \text{ or } \widetilde{x} > \overline{x}_i\} + 1 - \alpha) \cdot \widetilde{x} =$$

$$\alpha \cdot \sum_{i:s_j \geq \overline{x}_i} \overline{x}_i + \alpha \cdot \sum_{i:s_{j+1} \leq \underline{x}_i} \underline{x}_i + (1 - \alpha) \cdot \sum_{i:s_{j+1} \leq \widetilde{x}_i} \overline{x}_i + (1 - \alpha) \cdot \sum_{i:s_j \geq \widetilde{x}_i} \underline{x}_i.$$

From this equation, we can easily find the desired expression for the value $\widetilde{x}$ at which the derivative is 0.

Thus, we arrive at the following algorithm.

2.2.4 Resulting Algorithm

First, for each interval $[\underline{x}_i, \overline{x}_i]$, we compute its midpoint $\widetilde{x}_i = \dfrac{\underline{x}_i + \overline{x}_i}{2}$. Then, we sort the $3n$ values $\underline{x}_i$, $\overline{x}_i$, and $\widetilde{x}_i$ into an increasing sequence $s_1 \leq s_2 \leq \cdots \leq s_{3n}$. To cover the whole real line, to these values, we add $s_0 = -\infty$ and $s_{3n+1} = +\infty$.

We compute the value of the objective function (2.2.1) on each of the endpoints $s_1, \ldots, s_{3n}$. Then, for each interval (s_i, s_{j+1}), we compute the value

$$\widetilde{x} = \frac{\alpha \cdot \displaystyle\sum_{i:s_j \geq \overline{x}_i} \overline{x}_i + \alpha \cdot \displaystyle\sum_{i:s_{j+1} \leq \underline{x}_i} \underline{x}_i + (1 - \alpha) \cdot \displaystyle\sum_{i:s_{j+1} \leq \widetilde{x}_i} \overline{x}_i + (1 - \alpha) \cdot \displaystyle\sum_{i:s_j \geq \widetilde{x}_i} \underline{x}_i}{\alpha \cdot \#\{i : \widetilde{x} < \underline{x}_i \text{ or } \widetilde{x} > \overline{x}_i\} + 1 - \alpha}.$$

If the resulting value $\widetilde{x}$ is within the interval (s_i, s_{j+1}), we compute the value of the objective function (2.2.1) corresponding to this $\widetilde{x}$.

After that, out of all the values $\widetilde{x}$ for which we have computed the value of the objective function (2.2.1), we return the value $\widetilde{x}$ for which objective function $D(\widetilde{x})$ was the smallest.

What is the computational complexity of this algorithm. Sorting $3n = O(n)$ values $\underline{x}_i$, $\overline{x}_i$, and $\widetilde{x}_i$ takes time $O(n \cdot \ln(n))$.

Computing each value $D(\widetilde{x})$ of the objective function requires $O(n)$ computational steps. We compute $D(\widetilde{x})$ for $3n$ endpoints and for $\leq 3n + 1$ values at which the derivative is 0 at each of the intervals (s_j, s_{j+1}) – for the total of $O(n)$ values.

Thus, overall, we need $O(n \cdot \ln(n)) + O(n) \cdot O(n) = O(n^2)$ computation steps. Hence, our algorithm runs in quadratic time.

2.3 How to Get More Accurate Estimates – By Properly Taking Model Inaccuracy into Account

In engineering design, it is important to guarantee that the values of certain quantities such as stress level, noise level, vibration level, etc., stay below a certain threshold in all possible situations, i.e., for all possible combinations of the corresponding internal and external parameters. Usually, the number of possible combinations is so large that it is not possible to physically test the system for all these combinations. Instead, we form a computer model of the system, and test this model. In this testing, we need to take into account that the computer models are usually approximate. In this section, we show that the existing techniques for taking model uncertainty into account overestimate the uncertainty of the results. We also show how we can get more accurate estimates.

Results described in this section first appeared in [35, 36].

2.3.1 What if We Take into Account Model Inaccuracy

Models are rarely exact. Engineering systems are usually complex. As a result, it is rarely possible to find explicit expressions for y as a function of the parameters $x_1, \ldots, x_n$. Usually, we have some approximate computations. For example, if y is obtained by solving a system of partial differential equations, we use, e.g., the Finite Element method to find the approximate solution and thus, the approximate value of the quantity y.

How model inaccuracy is usually described. In most practical situations, at best, we know the upper bound ε on the accuracy of the computational model. In such cases, for each tuple of parameters $x_1, \ldots, x_n$, once we apply the computational model and get the value $F(x_1, \ldots, x_n)$, the actual (unknown) value $f(x_1, \ldots, x_n)$ of the quantity y satisfies the inequality

$$|F(x_1, \ldots, x_n) - f(x_1, \ldots, x_n)| \le \varepsilon. \tag{2.3.1}$$

How this model inaccuracy affects the above checking algorithms: analysis of the problem. Let us start with the formula (1.7.3). This formula assumes that we know the exact values of $\widetilde{y} = f(\widetilde{x}_1, \ldots, \widetilde{x}_n)$ and

$$y_i \overset{\text{def}}{=} f(\widetilde{x}_1, \ldots, \widetilde{x}_{i-1}, \widetilde{x}_i + \Delta_i, \widetilde{x}_{i+1}, \ldots, \widetilde{x}_n).$$

Instead, we know the values

$$\widetilde{Y} \overset{\text{def}}{=} F(\widetilde{x}_1, \ldots, \widetilde{x}_n) \tag{2.3.2}$$

and

$$Y_i \overset{\text{def}}{=} F(\widetilde{x}_1, \ldots, \widetilde{x}_{i-1}, \widetilde{x}_i + \Delta_i, \widetilde{x}_{i+1}, \ldots, \widetilde{x}_n) \tag{2.3.3}$$

which are ε-close to the values $\widetilde{y}$ and y_i. We can apply the formula (1.7.3) to these approximate values, and get the upper and lower bounds:

$$\overline{Y} = \widetilde{y} + \sum_{i=1}^{n} |Y_i - \widetilde{y}|; \quad \underline{Y} = \widetilde{y} - \sum_{i=1}^{n} |Y_i - \widetilde{y}|. \tag{2.3.4}$$

Here, $|\widetilde{Y} - \widetilde{y}| \leq \varepsilon$ and $|Y_i - y_i| \leq \varepsilon$, hence $|(Y_i - \widetilde{Y}) - (y_i - \widetilde{y})| \leq 2\varepsilon$ and $||Y_i - \widetilde{Y}| - |y_i - \widetilde{y}|| \leq 2\varepsilon$. By adding up all these inequalities, we conclude that

$$|\overline{y} - \overline{Y}| \leq (2n + 1) \cdot \varepsilon \text{ and } \left|\underline{y} - \underline{Y}\right| \leq (2n + 1) \cdot \varepsilon. \tag{2.3.5}$$

Thus, the only information that we have about the desired upper bound $\overline{y}$ is that $\overline{y} \leq B$, where

$$B \overset{\text{def}}{=} \overline{Y} + (2n + 1) \cdot \varepsilon. \tag{2.3.6}$$

Hence, we arrive at the following method.

Resulting method. We know:

- an algorithm $F(x_1, \ldots, x_n)$ that, given the values of the parameters $x_1, \ldots, x_n$, computes the value of the quantity y with a known accuracy ε;
- for each parameter x_i, we know its nominal value $\widetilde{x}_i$ and the largest possible deviation Δ_i from this nominal value.

Based on this information, we need to find an enclosure for the range of all the values values $f(x_1, \ldots, x_n)$ for all possible combinations of values x_i from the corresponding intervals

$$[\widetilde{x}_i - \Delta_i, \widetilde{x}_i + \Delta_i].$$

We can perform this computation as follows:

(1) first, we apply the algorithm F to compute the value

$$\widetilde{Y} = F(\widetilde{x}_1, \ldots, \widetilde{x}_n); \tag{2.3.7}$$

(2) then, for each i from 1 to n, we apply the algorithm F to compute the value

$$Y_i = F(\widetilde{x}_1, \ldots, \widetilde{x}_{i-1}, \widetilde{x}_i + \Delta_i, \widetilde{x}_{i+1}, \ldots, \widetilde{x}_n); \tag{2.3.8}$$

(3) after that, we compute the interval $\left[\widetilde{Y} - \Delta, \widetilde{Y} + \Delta\right]$, where

$$\Delta = \sum_{i=1}^{n} \left| Y_i - \widetilde{Y} \right| + (2n + 1) \cdot \varepsilon. \tag{2.3.9}$$

Comment 1. Please note that, in contrast to the case of the exact model, if, e.g., the upper bound $\widetilde{Y} + \Delta$ is larger than a given threshold t, this does not necessarily mean that the corresponding specification $y \leq t$ is not satisfied: maybe it is satisfied, but we cannot check that since we only know approximate values of y.

Comment 2. Similar bounds can be found for the estimates based on the Cauchy distribution.

Comment 3. The above estimate B is not the best that we can get, but it has been proven that computing the best estimate would require un-realistic exponential time [29, 37] – i.e., time which grows as 2^s with the size s of the input; thus, when we only consider feasible algorithms, overestimation is inevitable.

Comment 4. Similar to the methods described in Chap. 1, instead of directly applying the algorithm F to the modified tuples, we can, wherever appropriate, use the above-mentioned local sensitivity analysis technique.

Problem. When n is large, then, even for reasonably small inaccuracy ε, the value $(2n + 1) \cdot \varepsilon$ is large.

In this section, we show how we can get better estimates for the difference between the desired bounds $\widetilde{y}$ and $\underline{y}$ and the computed bounds $\overline{Y}$ and $\underline{Y}$.

2.3.2   How to Get Better Estimates

Main idea. As we have mentioned earlier, usually, we know the partial differential equations that describe the engineering system. Model inaccuracy comes from the fact that we do not have an analytical solution to this system of equations, so we have to use numerical (approximate) methods.

Usual numerical methods for solving systems of partial differential equations involve discretization of space – e.g., the use of Finite Element Methods.

Strictly speaking, the resulting inaccuracy is deterministic. However, in most cases, for all practical purposes, this inaccuracy can be viewed as random: when we select a different combination of parameters, we get an unrelated value of discretization-based inaccuracy.

In other words, we can view the differences

$$F(x_1, \ldots, x_n) - f(x_1, \ldots, x_n) \tag{2.3.10}$$

corresponding to different tuples $(x_1, \ldots, x_n)$ as independent random variables. In particular, this means that the differences $\overline{Y} - \widetilde{y}$ and $Y_i - y_i$ are independent random variables.

Technical details. What is a probability distribution for these random variables?

All we know about each of these variables is that its values are located somewhere in the interval $[-\varepsilon, \varepsilon]$. We do not have any reason to assume that some values from this interval are more probable than others, so it is reasonable to assume that all the values are equally probable, i.e., that we have a uniform distribution on this interval.

For this uniform distribution, the mean is 0, and the standard deviation is

$$\sigma = \frac{\varepsilon}{\sqrt{3}}.$$

Auxiliary idea: how to get a better estimate for $\tilde{y}$. In our main algorithm, we apply the computational model F to $n + 1$ different tuples. What we suggest it to apply it to one more tuple (making it $n + 2$ tuples), namely, computing an approximation

$$M \stackrel{\text{def}}{=} F(\tilde{x}_1 - \Delta_1, \ldots, \tilde{x}_n - \Delta_n) \qquad (2.3.11)$$

to the value

$$m \stackrel{\text{def}}{=} f(\tilde{x}_1 - \Delta_1, \ldots, \tilde{x}_n - \Delta_n). \qquad (2.3.12)$$

In the linearized case, one can easily check that

$$\tilde{y} + \sum_{i=1}^{n} y_i + m = (n + 2) \cdot \tilde{y}, \qquad (2.3.13)$$

i.e.,

$$\tilde{y} = \frac{1}{n + 2} \cdot \left(\tilde{y} + \sum_{i=1}^{n} y_i + m \right). \qquad (2.3.14)$$

Thus, we can use the following formula to come up with a new estimate $\tilde{Y}_{\text{new}}$ for $\tilde{y}$:

$$\tilde{Y}_{\text{new}} = \frac{1}{n + 2} \cdot \left(\tilde{Y} + \sum_{i=1}^{n} Y_i + m \right). \qquad (2.3.15)$$

For the differences $\Delta \tilde{y}_{\text{new}} \stackrel{\text{def}}{=} \tilde{Y}_{\text{new}} - \tilde{y}$, $\Delta \tilde{y} \stackrel{\text{def}}{=} \tilde{y} - \tilde{y}$, $\Delta y_i \stackrel{\text{def}}{=} Y_i - y_i$, and $\Delta m \stackrel{\text{def}}{=} M - m$, we have the following formula:

$$\Delta \tilde{y}_{\text{new}} = \frac{1}{n + 2} \cdot \left(\Delta \tilde{y} + \sum_{i=1}^{n} \Delta y_i + \Delta m \right). \qquad (2.3.16)$$

The left-hand side is the arithmetic average of $n + 2$ independent identically distributed random variables, with mean 0 and variance $\sigma^2 = \dfrac{\varepsilon^2}{3}$. Hence (see, e.g.,

[97]), the mean of this average $\Delta \widetilde{y}_{\text{new}}$ is the average of the means, i.e., 0, and the variance is equal to $\sigma^2 = \dfrac{\varepsilon^2}{3 \cdot (n+2)} \ll \dfrac{\varepsilon^2}{3} = \sigma^2[\Delta \widetilde{y}]$.

Thus, this average $\widetilde{Y}_{\text{new}}$ is a more accurate estimation of the quantity $\widetilde{y}$ than $\widetilde{Y}$.

Let us use this better estimate for $\widetilde{y}$ when estimating the range of y. Since the average $\widetilde{Y}_{\text{new}}$ is a more accurate estimation of the quantity $\widetilde{y}$ than $\widetilde{Y}$, let us use this average instead of $\widetilde{y}$ when estimating $\overline{Y}$ and $\underline{Y}$. In other words, instead of the estimate (2.3.9), let us use a new estimate $[\overline{Y}_{\text{new}} - \Delta_{\text{new}}, \overline{Y}_{\text{new}} + \Delta_{\text{new}}]$, where

$$\Delta_{\text{new}} = \sum_{i=1}^{n} \left| Y_i - \widetilde{Y}_{\text{new}} \right|. \tag{2.3.17}$$

Let us estimate the accuracy of this new approximation.

The formula (1.7.3) can be described in the following equivalent form:

$$\overline{y} = \widetilde{y} + \sum_{i=1}^{n} s_i \cdot (y_i - \widetilde{y}) = \left(1 - \sum_{i=1}^{n} s_i \right) \cdot \widetilde{y} + \sum_{i=1}^{n} s_i \cdot y_i, \tag{2.3.18}$$

where $s_i \in \{-1, 1\}$ are the signs of the differences $y_i - \widetilde{y}$.

Similarly, we get

$$\overline{Y}_{\text{new}} = \left(1 - \sum_{i=1}^{n} s_i \right) \cdot \widetilde{Y}_{\text{new}} + \sum_{i=1}^{n} s_i \cdot Y_i. \tag{2.3.19}$$

Thus, e.g., for the difference $\Delta \overline{y} \overset{\text{def}}{=} \overline{Y}_{\text{new}} - \overline{y}$, we have

$$\Delta \overline{y}_{\text{new}} = \left(1 - \sum_{i=1}^{n} s_i \right) \cdot \Delta \widetilde{y}_{\text{new}} + \sum_{i=1}^{n} s_i \cdot \Delta y_i. \tag{2.3.20}$$

Here, the differences $\Delta \widetilde{y}_{\text{new}}$ and Δy_i are independent random variables. According to the Central Limit Theorem (see, e.g., [97]), for large n, the distribution of a linear combination of many independent random variables is close to Gaussian. The mean of the resulting distribution is the linear combination of the means, thus equal to 0.

The variance of a linear combination $\sum_i k_i \cdot \eta_i$ of independent random variables η_i with variances σ_i^2 is equal to $\sum_i k_i^2 \cdot \sigma_i^2$. Thus, in our case, the variance σ^2 of the difference $\Delta \overline{y}$ is equal to

$$\sigma^2 = \left(1 - \sum_{i=1}^{n} s_i \right)^2 \cdot \frac{\varepsilon^2}{3 \cdot (n+2)} + \sum_{i=1}^{n} \frac{\varepsilon^2}{3}. \tag{2.3.21}$$

Here, since $|s_i| \leq 1$, we have $\left| 1 - \sum_{i=1}^{n} s_i \right| \leq n + 1$, so (2.3.21) implies that

$$\sigma^2 \leq \frac{\varepsilon^2}{3} \cdot \left(\frac{(n+1)^2}{n+2} + n \right). \tag{2.3.22}$$

Here, $\dfrac{(n+1)^2}{n+2} \leq \dfrac{(n+1)^2}{n+1} = n + 1$, hence

$$\sigma^2 \leq \frac{\varepsilon^2}{3} \cdot (2n + 1). \tag{2.3.23}$$

For a normal distribution, with almost complete certainty, all the values are concentrated within k_0 standard deviations away from the mean: within 2σ with confidence 0.95, within 3σ with degree of confidence 0.999, within 6σ with degree of confidence $1 - 10^{-8}$. Thus, we can safely conclude that

$$\overline{y} \leq \overline{Y}_{new} + k_0 \cdot \sigma \leq \overline{Y}_{new} + k_0 \cdot \frac{\varepsilon}{\sqrt{3}} \cdot \sqrt{2n + 1}. \tag{2.3.24}$$

Here, inaccuracy grows as $\sqrt{2n + 1}$, which is much better than in the traditional approach, where it grows proportionally to $2n + 1$ – and we achieve this drastic reduction of the overestimation, basically by using one more run of the program F in addition to the previously used $n + 1$ runs.

Similar estimates can be made for the lower bound $\underline{y}$. So, we arrive at the following method.

Resulting method. We know:

- an algorithm $F(x_1, \ldots, x_n)$ that, given the values of the parameters $x_1, \ldots, x_n$, computes the value of the quantity y with a known accuracy ε;
- for each parameter x_i, we know its nominal value $\widetilde{x}_i$ and the largest possible deviation Δ_i from this nominal value.

Based on this information, we need to estimate the range of $y = f(x_1, \ldots, x_n)$ when x_i are from the corresponding intervals $[\widetilde{x}_i - \Delta_i, \widetilde{x}_i + \Delta_i]$.

We can estimate this range as follows:

(1) first, we apply the algorithm F to compute the value

$$\widetilde{Y} = F(\widetilde{x}_1, \ldots, \widetilde{x}_n); \tag{2.3.25}$$

(2) then, for each i from 1 to n, we apply the algorithm F to compute the value

$$Y_i = F(\widetilde{x}_1, \ldots, \widetilde{x}_{i-1}, \widetilde{x}_i + \Delta_i, \widetilde{x}_{i+1}, \ldots, \widetilde{x}_n); \tag{2.3.26}$$

(3) then, we compute

$$M = F(\tilde{x}_1 - \Delta_1, \ldots, \tilde{x}_n - \Delta_n);$$ (2.3.27)

(4) we compute

$$\tilde{Y}_{\text{new}} = \frac{1}{n+2} \cdot \left(\tilde{Y} + \sum_{i=1}^{n} Y_i + M \right);$$ (2.3.28)

(5) finally, we compute $\left[\tilde{Y}_{\text{new}} - \tilde{\Delta}_{\text{new}}, \tilde{Y}_{\text{new}} + \tilde{\Delta}_{\text{new}} \right]$, where

$$\tilde{\Delta}_{\text{new}} = \sum_{i=1}^{n} \left| Y_i - \tilde{Y}_{\text{new}} \right| + k_0 \cdot \sqrt{2n+1} \cdot \frac{\varepsilon}{\sqrt{3}},$$ (2.3.29)

where k_0 depends on the level of confidence that we can achieve.

Comment. For the method based on Cauchy distribution, similarly, after computing $\tilde{Y} = F(\tilde{x}_1, \ldots, \tilde{x}_n)$ and $Y^{(k)} = F(\tilde{x}_1 + \eta_1^{(k)}, \ldots, \tilde{x}_n + \eta_n^{(k)})$, we can compute the improved estimate $\tilde{Y}_{\text{new}}$ for $\tilde{y}$ as

$$\tilde{Y}_{\text{new}} = \frac{1}{N+1} \cdot \left(\tilde{Y} + \sum_{k=1}^{N} Y^{(k)} \right),$$ (2.3.30)

and estimate the parameter Δ based on the more accurate differences

$$\Delta Y_{\text{new}}^{(k)} = Y^{(k)} - \tilde{Y}_{\text{new}}.$$

Experimental testing. We tested our approach on the example of the seismic inverse problem in geophysics, where we need to reconstruct the velocity of sound at different spatial locations and at different depths based on the times that it takes for a seismic signal to get from the set-up explosion to different seismic stations. In this reconstruction, we used (a somewhat improved version of) the finite element technique that was originated by John Hole [24]; the resulting techniques are described in [4, 33, 67].

According to Hole's algorithm, we divide the 3-D volume of interest (in which we want to find the corresponding velocities) into a rectangular 3-D grid of N small cubic cells. We assume that the velocity is constant within each cube; the value of velocity in the jth cube is denoted by v_j. Each observation j means that we know the time t_j that it took the seismic wave to go from the site of the corresponding explosion to the location of the observing sensor.

This algorithm is iterative. We start with the first-approximation model of the Earth, namely, with geology-motivated approximate values $v_i^{(1)}$. At each iteration k, we start with the values $v_i^{(k)}$ and produce the next approximation $v_i^{(k+1)}$ as follows.

First, based on the latest approximation $v_i^{(k)}$, we simulate how the seismic waves propagate from the explosion site to the sensor locations. In the cube that contains the explosion site, the seismic signal propagates in all directions. When the signal's trajectory approaches the border between the two cubes i and i', the direction of the seismic wave changes in accordance with the Snell's law $\dfrac{\sin(\theta_i)}{\sin(\theta_{i'})} = \dfrac{v_i^{(k)}}{v_{i'}^{(k)}}$, where θ_i is the angle between the seismic wave's trajectory in the ith cube and the vector orthogonal to the plane separating the two cubes. Snell's law enables us to find the trajectory's direction in the next cube i'. Once the way reaches the location of the sensor, we can estimate the travel time as $t_j^{(k)} = \sum\limits_i \dfrac{\ell_{ji}}{v_i^{(k)}}$, where the sum is taken over all the cubes through which this trajectory passes, and ℓ_{ji} is the length of the part of the trajectory that lies in the ith cube.

Each predicted value $t_j^{(k)}$ is, in general, different from the observed value t_j. To compensate for this difference, the velocity model $v_i^{(k)}$ is corrected: namely, the inverse value $s_i^{(k)} \stackrel{\text{def}}{=} \dfrac{1}{v_i^{(k)}}$ is replaced with an updated value

$$s_i^{(k+1)} = s_i^{(k)} + \frac{1}{n_i} \cdot \sum_j \frac{t_j - t_j^{(k)}}{L_j}, \tag{2.3.31}$$

where the sum is taken over all trajectories that pass through the ith cube, n_i is the overall number of such trajectories, and $L_j = \sum\limits_i \ell_{ji}$ is the overall length of the jth trajectory.

Iterations stop when the process converges; for example, it is reasonable to stop the process when the velocity models obtained on two consecutive iterations becomes close:

$$\sum_i \left(v_i^{(k+1)} - v_i^{(k)} \right)^2 \leq \varepsilon \tag{2.3.32}$$

for some small value $\varepsilon > 0$. The quality of the resulting solution can be gauged by how well the predicted travel times $t_i^{(k)}$ match the observations; usually, by the root mean square (rms) approximation error $\sqrt{\dfrac{1}{N} \cdot \sum\limits_i \left(t_i^{(k)} - t_i \right)^2}$.

In this problem, there are two sources of uncertainty. The first is the uncertainty with which we can measure each travel time t_j. The travel time is the difference between the time when the signal arrives at the sensor location and the time of the artificially set explosion. The explosion time is known with a very high accuracy, but the arrival time is not. In the ideal situation, when the only seismic signal comes from the our explosion, we could exactly pinpoint the arrival time as the time when the sensor starts detecting a signal. In real life, there is always a background noise, so we can only determine the arrival time with some accuracy.

The second source of uncertainty comes from the fact that our discrete model is only an approximate description of the continuous real Earth.

In [4, 33, 67], we used the formula (1.7.3) and the Cauchy-based techniques to estimate how the measurement uncertainty affects the results of data processing. To test the method described in this section, we used the above formulas to compute the new value $\widetilde{y}_{new}$. These new values indeed lead to a better fit with data than the original values $\widetilde{y}$: in our 16 experiments, we only in one case, the rms approximation error decreased, on average, by 15%. It should also be mentioned that only in one case the rms approximation error increased (and not much, only by 7%); in all other 15 cases, the rms approximation error decreased.

2.3.3   Can We Further Improve the Accuracy?

How to improve the accuracy: a straightforward approach. As we have mentioned, the inaccuracy $F \neq f$ is caused by the fact that we are using a Finite Element method with a finite size elements. In the traditional Finite Element method, when we assume that the values of each quantity within each element are constant, this inaccuracy comes from the fact that we ignore the difference between the values of the corresponding parameters within each element. For elements of linear size h, this inaccuracy Δy is proportional to $y' \cdot h$, where y' is the spatial derivative of y. In other words, the inaccuracy is proportional to the linear size h.

A straightforward way to improve the accuracy is to decrease h. For example, if we reduce h to $\dfrac{h}{2}$, then we decrease the resulting model inaccuracy ε to $\dfrac{\varepsilon}{2}$.

This decrease requires more computations. The number of computations is, crudely speaking, proportional to the number of nodes. Since the elements fill the original area, and each element has volume h^3, the number of such elements is proportional to h^{-3}.

So, if we go from the original value h to the smaller value h', then we increase the number of computations by a factor of

$$K \stackrel{\text{def}}{=} \frac{h^3}{(h')^3}.   \qquad (2.3.33)$$

This leads to decreasing the inaccuracy by a factor of $\dfrac{h}{h'}$, which is equal to $\sqrt[3]{K}$.

For example, in this straightforward approach, if we want to decrease the accuracy in half $\left(\sqrt[3]{K} = 2\right)$, we will have to increase the number of computation steps by a factor of $K = 8$.

An alternative approach: description. An alternative approach is as follows. We select K small vectors $\left(\Delta_1^{(k)}, \ldots, \Delta_n^{(k)} \right), 1 \le k \le K$, which add up to 0. For example, we can arbitrarily select the first $K - 1$ vectors and take $\Delta x_i^{(K)} = - \sum_{k=1}^{K-1} \Delta_i^{(k)}$.

Then, every time we need to estimate the value $f(x_1, \ldots, x_n)$, instead of computing $F(x_1, \ldots, x_n)$, we compute the average

$$F_K(x_1, \ldots, x_n) = \frac{1}{K} \cdot \sum_{k=1}^{K} F\left(x_1 + \Delta_1^{(k)}, \ldots, x_n + \Delta_n^{(k)} \right). \tag{2.3.34}$$

Why this approach decreases inaccuracy. We know that

$$F(x_1 + \Delta x_1, \ldots, x_n + \Delta x_n) = f(x_1 + \Delta x_1, \ldots, x_n + \Delta x_n) + \delta y,$$

where, in the small vicinity of the original tuple $(x_1, \ldots, x_n)$, the expression $f(x_1 + \Delta x_1, \ldots, x_n + \Delta x_n)$ is linear, and the differences δy are independent random variables with zero mean.

Thus, we have

$$F_K(x_1, \ldots, x_n) = \frac{1}{K} \cdot \sum_{k=1}^{K} f\left(x_1 + \Delta_1^{(k)}, \ldots, x_n + \Delta_n^{(k)} \right) + \frac{1}{K} \cdot \sum_{k=1}^{K} \delta y^{(k)}. \tag{2.3.35}$$

Due to linearity and the fact that $\sum_{k=1}^{K} \Delta_i^{(k)} = 0$, the first average in (2.3.35) is equal to $f(x_1, \ldots, x_n)$. The second average is the average of K independent identically distributed random variables, and we have already recalled that this averaging decreases the inaccuracy by a factor of $\sqrt{K}$.

Thus, in this alternative approach, we increase the amount of computations by a factor of K, and as a result, we decrease the inaccuracy by a factor of $\sqrt{K}$.

The new approach is better than the straightforward one. In general, $\sqrt{K} > \sqrt[3]{K}$. Thus, with the same increase in computation time, the new method provides a better improvement in accuracy than the straightforward approach.

Comment. The above computations only refer to the traditional Finite Element approach, when we approximate each quantity within each element by a *constant*. In many real-life situations, it is useful to approximate each quantity within each element not by a constant, but rather by a *polynomial* of a given order (see, e.g., [99]): by a linear function, by a quadratic function, etc. In this case, for each element size h, we have smaller approximation error but larger amount of computations. It is desirable to extend the above analysis to such techniques as well.

2.4 How to Gauge the Accuracy of Fuzzy Control Recommendations: A Simple Idea

Fuzzy control is based on approximate expert information, so its recommendations are also approximate. However, the traditional fuzzy control algorithms do not tell us how accurate are these recommendations. In contrast, for the probabilistic uncertainty, there is a natural measure of accuracy: namely, the standard deviation. In this paper, we show how to extend this idea from the probabilistic to fuzzy uncertainty and thus, to come up with a reasonable way to gauge the accuracy of fuzzy control recommendations.

Results of this section first appeared in [50].

2.4.1 Formulation of the Problem

Need to gauge accuracy of fuzzy recommendations. Fuzzy logic (see, e.g., [28, 61, 108]) has been successfully applied to many different application areas.

For example, in control – one of the main applications of fuzzy techniques – fuzzy techniques enable us to generate the control value appropriate for a given situation.

A natural question is: with what accuracy do we need to implement this recommendation? In many applications, this is an important question: it is often much easier to implement the control value approximately, by using a simple approximate actuator, but maybe a more accurate actuator is needed? To answer this question, we must have a natural way to gauge the accuracy of the corresponding recommendations.

Such gauging is possible for probabilistic uncertainty. In a similar case of probabilistic uncertainty, there is such a natural way to gauge the accuracy; see, e.g., [97].

Namely, probabilistic uncertainty means that instead of the exact value x, we only know a probability distribution – which can be described, e.g., by the probability density $\rho(x)$. In this situation, if we need to select a single value x, a natural idea is to select, e.g., the mean value $\overline{x} = \int x \cdot \rho(x)\, dx$.

A natural measure of accuracy of this means is the mean square deviation from the mean, known as the standard deviation:

$$\sigma \stackrel{\text{def}}{=} \sqrt{\int (x - \overline{x})^2 \, dx}.$$

What we do in this section. In this section, we provide a similar way to gauge the accuracy of fuzzy recommendations, i.e., a recommendations in which, instead of using a probability density function $\rho(x)$, we start with a membership function $\mu(x)$.

2.4.2 Main Idea

How we elicit fuzzy degrees: a brief reminder. To explain our idea, let us recall how fuzzy degrees $\mu(x)$ corresponding to different values x are elicited in the first place.

At first glance, the situation may look straightforward: for each possible value x of the corresponding quantity, we ask the expert to mark, on a scale from 0 to 1, his/her degree of confidence that x satisfies the given property. For example, if we are eliciting the membership function describing smallness, we ask the expert to specify the degree to which the value x is small.

In some cases, this is all we need. However, in many other cases, we get a *non-normalized* membership function, for which the largest value $\mu(x)$ is smaller than 1. Most fuzzy techniques assume that the membership function is normalized. So, after the elicitation, we sometimes need to perform an additional step to get an easy-to-process membership function: namely, we *normalize* the original values $\mu(x)$ by dividing them by the largest of the values $\mu(y)$. Thus, we get the function

$$\mu_n(x) \overset{\text{def}}{=} \frac{\mu(x)}{\max_y \mu(y)}.$$

Sometimes, the original fuzzy degrees come from subjective probabilities. Sometimes, the experts have some subjective probabilities assigned to different values x. In this case, when asked to indicate their degree of certainty, they may list the values of the corresponding probability density function $\rho(x)$.

This function is rarely normalized. After normalizing it, we get the membership function

$$\mu(x) = \frac{\rho(x)}{\max_y \rho(y)}. \tag{2.4.1}$$

Let us use this idea to gauge the accuracy of fuzzy recommendations. Formula (2.4.1) assigns, to each probability density function $\rho(x)$, an appropriate membership function $\mu(x)$. Vice versa, one can easily see if we know that the membership function $\mu(x)$ was obtained by normalizing some probability density function $\rho(x)$, then we can uniquely reconstruct this probability density function $\rho(x)$: namely, since $\mu(x) = c \cdot \rho(x)$ for some normalizing constant c, we thus have $\rho(x) = C \cdot \mu(x)$, for another constant $C = \dfrac{1}{c}$. So, all we need to find the probability density function is to find the coefficient C.

This coefficient can be easily found from the condition that the overall probability be 1, i.e., that $\int \rho(x)\,dx = 1$. Substituting $\rho(x) = C \cdot \mu(x)$ into this formula, we conclude that $C \cdot \int \mu(x)\,dx = 1$, thus $C = \dfrac{1}{\int \mu(y)\,dy}$ and therefore,

$$\rho(x) = C \cdot \mu(x) = \frac{\mu(x)}{\int \mu(y)\, dy}. \tag{2.4.2}$$

Our idea is then to use the probabilistic formulas corresponding to this artificial distribution.

This makes sense. Does this make sense? The probabilistic measure of accuracy is based on the assumption that we use the mean, but don't we use something else in fuzzy?

Actually, not really. The mean of the distribution (2.4.2) is

$$\overline{x} = \int x \cdot \rho(x)\, dx = \frac{\int x \cdot \mu(x)\, dx}{\int \mu(x)\, dx}.$$

This is exactly the centroid defuzzification – one of the main ways to transform the membership function into a single numerical control recommendation.

Since the above idea makes sense, let us use it to gauge the accuracy of the fuzzy control recommendation.

Resulting recommendation. For a given membership function $\mu(x)$, in addition to the result $\overline{x}$ of its centroid defuzzification, we should also generate, as a measure of the accuracy of this recommendation, the value σ which is defined by the following formula

$$\sigma^2 = \int (x - \overline{x})^2 \cdot \rho(x)\, dx = \frac{\int (x - \overline{x})^2 \cdot \mu(x)\, dx}{\int \mu(x)\, dx} =$$

$$\frac{\int x^2 \cdot \mu(x)\, dx}{\int \mu(x)\, dx} - \left(\frac{\int x \cdot \mu(x)\, dx}{\int \mu(x)\, dx} \right)^2. \tag{2.4.3}$$

2.4.3 But What Should We Do in the Interval-Valued Fuzzy Case?

But what do we do for type-2 fuzzy logic? For the above case of type-1 fuzzy logic, this is just a simple recommendation.

But what do we do if we use a more adequate way to describe uncertainty – namely, type-2 fuzzy logic? In this section, we consider the simplest case of type-2 fuzzy logic – the interval-valued fuzzy logic (see, e.g., [51, 52]), where for each possible value x of the corresponding quantity, we only know the interval $[\underline{\mu}(x), \overline{\mu}(x)]$ of possible value of degree of confidence $\mu(x)$?

Challenge. In this case, we have a challenge:

- just like for defuzzification, we need to find the range of possible values of $\overline{x}$ corresponding to different functions $\mu(x)$ from the given interval [51, 52],

- similarly, we need to find the range of possible values of σ^2 when each value $\mu(x)$ belongs to the corresponding interval.

Analysis of the problem. According to calculus, when the maximum of a function $f(z)$ on the interval $[\underline{z}, \overline{z}]$ is attained at some point $z_0 \in [\underline{z}, \overline{z}]$, then we have one of the three possible cases:

- we can have $z_0 \in (\underline{z}, \overline{z})$, in which case $\dfrac{df}{dz} = 0$ at this point z_0;
- we can have $z_0 = \underline{z}$, in this case, we must have $\dfrac{df}{dz} \leq 0$ at this point (otherwise, the function would increase even further when z increases, and so there would no maximum at $\underline{z}$), or
- we can have $z_0 = \overline{z}$, in which case $\dfrac{df}{dz} \geq 0$.

Similarly, when the minimum of a function $f(z)$ on the interval $[\underline{z}, \overline{z}]$ is attained at some point $z_0 \in [\underline{z}, \overline{z}]$, then we have one of the three possible cases:

- we can have $z_0 \in (\underline{z}, \overline{z})$, in which case $\dfrac{df}{dz} = 0$ at this point z_0;
- we can have $z_0 = \underline{z}$, in this case, we must have $\dfrac{df}{dz} \geq 0$ at this point, or
- we can have $z_0 = \overline{z}$, in which case $\dfrac{df}{dz} \leq 0$.

Let us apply this general idea to the dependence of the expression (2.4.3) on each value $\mu(a)$.

Here, taking into account that for $\int \mu(x)\, dx \approx \sum \mu(x_i) \cdot \Delta x_i$, we get

$$\frac{\partial(\int \mu(x)\, dx)}{\partial(\mu(a))} = \Delta x, \quad \frac{\partial(\int x \cdot \mu(x)\, dx)}{\partial(\mu(a))} = a \cdot \Delta x \text{ and}$$

$$\frac{\partial(\int x^2 \cdot \mu(x)\, dx)}{\partial(\mu(a))} = a^2 \cdot \Delta x.$$

Now, by using the usual rules for differentiating the ratio, for the composition, and for the square, we conclude that:

$$\frac{\partial(\sigma^2)}{\partial(\mu(a))} = \Delta x \cdot S(a),$$

where we denoted

$$S(a) \stackrel{\text{def}}{=} \frac{a^2}{\int \mu(x)\, dx} - \frac{\int x^2 \cdot \mu(x)\, dx}{\left(\int \mu(x)\, dx\right)^2} - 2 \cdot \overline{x} \cdot \left(\frac{a}{\int \mu(x)\, dx} - \frac{\int x \cdot \mu(x)\, dx}{\left(\int \mu(x)\, dx\right)^2}\right).$$

$$(2.4.4)$$

We are only interested in the sign of the derivative, so we can as well consider the sign of the expression $S(a)$ instead of the sign of the desired derivative $\dfrac{\partial(\sigma^2)}{\partial(\mu(a))}$.

Similar, the sign of the expression $S(a)$ is the same as the sign of the expression $s(a) \stackrel{\text{def}}{=} S(a) \cdot \int \mu(y)\,dy$ which has a simpler form

$$s(a) = a^2 - ((\overline{x})^2 + \sigma^2) - 2 \cdot \overline{x} \cdot (a - \overline{x}).$$

If we know the roots $\underline{x} < \overline{x}$ of this quadratic expression, we can conclude that this quadratic expression $s(a)$ is:

- positive when $a < \underline{x}$ and
- negative when $a > \overline{x}$.

Here, the value $a = \overline{x}$ is between $\underline{x}$ and $\overline{x}$, since for this value a, we have

$$s(\overline{x}) = -\sigma^2 < 0.$$

Thus, in accordance with the above fact from calculus:

- when $a < \underline{x}$ or $a > \overline{x}$, then to find the upper bound for σ^2, we must take $\mu(a) = \overline{\mu}(a)$ and to find the lower bound, we must take $\mu(a) = \underline{\mu}(a)$;
- when $\underline{x} < a < \overline{a}$, then, vice versa, we need to take $\mu(a) = \underline{\mu}(a)$ to find the upper bound for σ^2 and we must take $\mu(a) = \overline{\mu}(a)$ to find the lower bound.

This mathematical conclusion makes perfect sense: to get the largest standard deviation, we must concentrate the distribution as much as possible on values outside the mean, and to get the smallest possible standard deviation, we concentrate it as much as possible on values close to the mean.

Thus, we arrive at the following algorithm.

Resulting algorithm. For all possible values $\underline{x} < \overline{x}$, we use the formula (2.4.3) to compute the values $\sigma^2(\mu^-)$ and $\sigma^2(\mu^+)$ for the following two functions $\mu^-(x)$ and $\mu^+(x)$:

- $\mu^+(x) = \overline{\mu}(x)$ when $x < \underline{x}$ or $x > \overline{x}$, and $\mu^+(x) = \underline{\mu}(x)$ when $\underline{x} < x < \overline{x}$;
- $\mu^-(x) = \underline{\mu}(x)$ when $x < \underline{x}$ or $x > \overline{x}$, and $\mu^-(x) = \overline{\mu}(x)$ when $\underline{x} < x < \overline{x}$.

Then:

- as the upper bound for σ^2, we take the maximum of the values $\sigma^2(\mu^+)$ corresponding to different pairs $\underline{x} < \overline{x}$, and
- as the lower bound for σ^2, we take the minimum of the values $\sigma^2(\mu^-)$ corresponding to different pairs $\underline{x} < \overline{x}$.

Chapter 3
How to Speed Up Computations

In this chapter, we describe cases for which we can speed up computations. We start, in Sect. 3.1, with the case of fuzzy uncertainty under min t-norm, for which, as we mentioned in Chap. 1, data processing means using Zadeh's extension principle, or, equivalently, interval computations on α-cuts. In this case, processing different types of uncertainty separately can drastically speed up computations. For the important case of t-norms different from min, a similar speedup is described in Sect. 3.2. In Sect. 3.3, we describe a speedup for the case of probabilistic uncertainty. Finally, in Sect. 3.4, we speculate on the possibility of use quantum computing to further speed up data processing.

3.1 How to Speed Up Processing of Fuzzy Data

In many practical situations, we make predictions based on the measured and/or estimated values of different physical quantities. The accuracy of these predictions depends on the accuracy of the corresponding measurements and expert estimates. Often, for each quantity, there are several different sources of inaccuracy. Usually, to estimate the prediction accuracy, we first combine, for each input, inaccuracies from different sources into a single expression, and then use these expressions to estimate the prediction accuracy. In this section, on the example of fuzzy uncertainty, we show that it is often more computationally efficient to process different types of uncertainty separately, i.e., to estimate inaccuracies in the prediction result caused by different types of uncertainty, and only then combine these inaccuracies into a single estimate.

The results described in this section first appeared in [100].

3.1.1 Cases for Which a Speedup Is Possible: A Description

In this section, we describe the cases of fuzzy uncertainty for which the computations can be made faster.

Simplest case: when all fuzzy numbers are of the same type. Sometimes, all membership functions are "of the same type", i.e., they all have the form $\mu(z) = \mu_0(k \cdot z)$ for some fixed symmetric function $\mu_0(z)$.

For example, frequently, we consider symmetric triangular functions; all these functions can be obtained from the standard triangular function

$$\mu_0(z) = \max(1 - |z|, 0)$$

by using an appropriate constant $k > 0$.

In this case, we can speed up computations. Let us show that in this simple case, we can drastically reduce the computation time that is needed to compute the desired α-cuts $^\alpha\Delta$.

Indeed, let $[-^\alpha\Delta_0, {}^\alpha\Delta_0]$ denote an α-ut corresponding to the membership function $\mu_0(z)$. This means that the inequality $\mu_0(z) \geq \alpha$ is equivalent to $|z| \leq {}^\alpha\Delta_0$. Then, for the membership function $\mu(z) = \mu_0(k \cdot z)$, the inequality $\mu(z) \geq \alpha$ describing its α-cut is equivalent to $\mu_0(k \cdot z) \geq \alpha$, i.e., to to $k \cdot |z| \leq {}^\alpha\Delta_0$ and thus, $|z| \leq \dfrac{1}{k} \cdot {}^\alpha\Delta_0$. Hence, the half-widths of the corresponding α-cuts are equal to

$$^\alpha\Delta = \frac{1}{k} \cdot {}^\alpha\Delta_0. \tag{3.1.1}$$

This equality holds for all α, in particular, for $\alpha = 0$, when we get

$$^0\Delta = \frac{1}{k} \cdot {}^0\Delta_0. \tag{3.1.2}$$

By dividing (3.1.1) by (3.1.2), we conclude that

$$\frac{^\alpha\Delta}{^0\Delta} = f(\alpha), \tag{3.1.3}$$

where we denoted

$$f(\alpha) \stackrel{\text{def}}{=} \frac{^\alpha\Delta_0}{^0\Delta_0}. \tag{3.1.4}$$

For example, for a triangular membership function, we have

$$f(\alpha) = 1 - \alpha. \tag{3.1.5}$$

Thus, if we know the type of the membership function (and hence, the corresponding function $f(\alpha)$), and we know the 0-cut, then we can reconstruct all α-cuts as

$$^{\alpha}\Delta = f(\alpha) \cdot {}^{0}\Delta, \tag{3.1.6}$$

i.e., by simply multiplying the 0-cuts by an appropriate factor $f(\alpha)$.

So, if all the membership functions $\mu_i(\Delta x_i)$ and $\mu_m(\Delta m)$ are of the same type, then, for every α, we have $^{\alpha}\Delta_i = f(\alpha) \cdot {}^{0}\Delta_i$. Substituting these expressions into the formula (3.1.1), we conclude that

$$^{\alpha}\Delta = \sum_{i=1}^{n} |c_i| \cdot f(\alpha) \cdot {}^{0}\Delta_i = f(\alpha) \cdot \sum_{i=1}^{n} |c_i| \cdot {}^{0}\Delta_i, \tag{3.1.7}$$

i.e., that

$$^{\alpha}\Delta = f(\alpha) \cdot {}^{0}\Delta. \tag{3.1.8}$$

Thus, if all the membership functions are of the same type $\mu_0(z)$, there is no need to apply the formula (3.1.7) eleven times: it is sufficient to compute it only once, e.g., for $\alpha = 0$. To find all other values $^{\alpha}\Delta$, we can then simply multiply the resulting value $^{0}\Delta$ by the factors $f(\alpha)$ corresponding to the type $\mu_0(z)$.

A more general case. A more general case is when we have a list of T different types of uncertainty – i.e., types of membership functions – and each approximation error Δx_i consists of $\le T$ components of the corresponding type. In other words, for each i, we have

$$\Delta x_i = \sum_{t=1}^{T} \Delta x_{i,t}, \tag{3.1.9}$$

where $\Delta x_{i,t}$ are uncertainties of the tth type, and we know the corresponding membership functions $\mu_{i,t}(\Delta x_{i,t})$.

For example, type $t = 1$ may correspond to intervals (which are, of course, a particular case of fuzzy uncertainty), type $t = 2$ may correspond to triangular membership functions, etc.

How this case is processed now.

- First, we use the known membership functions $\mu_{i,t}(\Delta x_{i,t})$ to find the memberships functions $\mu_i(\Delta x_i)$ that correspond to the sums (3.1.9).
- Then, we use these membership functions to compute the desired membership function $\mu(\Delta y)$.

On the second stage, we apply the formula (1.8.4) eleven times.

3.1.2 Main Idea of This Section

Main idea. As we have mentioned, at present, to find the membership function for Δy, we use the formula (1.4.5), in which each of the terms Δx_i is computed by using the formula (3.1.9).

A natural alternative idea is:

• to substitute the expressions (3.1.9) into the formula (1.4.5), and then
• to regroup the resulting terms by combining all the components of the same type t.

Technical details. Substituting the expressions (3.1.9) into the formula (1.8.4), we conclude that

$$\Delta y = \sum_{i=1}^{n} c_i \cdot \left(\sum_{t=1}^{T} \Delta x_{i,t} \right). \tag{3.1.10}$$

Now, grouping together all the terms corresponding to each type t, we conclude that

$$\Delta y = \sum_{t=1}^{T} \Delta y_t, \tag{3.1.11}$$

where

$$\Delta y_t \stackrel{\text{def}}{=} \sum_{i=1}^{n} c_i \cdot \Delta x_{i,t}. \tag{3.1.12}$$

This representation suggests the following new algorithm.

New algorithm: idea. For each t, since we are combining membership functions of the same type, computing these membership functions requires a single application of the formula (1.8.4), to compute the value $^0\Delta_t$ corresponding to $\alpha = 0$. The values corresponding to other values α, we simply multiply this value $^0\Delta_t$ by the coefficients $f_t(\alpha)$ corresponding to membership functions of type t.

Then, we add the resulting membership functions – by adding the corresponding α-cuts. Let us describe the resulting algorithm in detail.

New algorithm: in detail. We start with the values $^0\Delta_{i,t}$ for which the corresponding symmetric intervals $[-^0\Delta_{i,t}, {}^0\Delta_{i,t}]$ describe the 0-cuts of the corresponding membership functions $\mu_{i,t}(\Delta x_{i,t})$.

Based on these 0-cuts, we compute, for each type t, the values

$$^0\Delta_t = \sum_{i=1}^{n} |c_i| \cdot {}^0\Delta_{i,t}. \tag{3.1.13}$$

Then, for $\alpha = 0$, $\alpha = 0.1$, ..., and for $\alpha = 1.0$, we compute the values

$$^\alpha\Delta_t = f_t(\alpha) \cdot {}^0\Delta_t, \tag{3.1.14}$$

where the function $f_t(\alpha)$ is known for each type t. Finally, we add up α-cuts corresponding to different types t, to come up with the expression

$$^{\alpha}\Delta = \sum_{t=1}^{T} {}^{\alpha}\Delta_t. \qquad (3.1.15)$$

Comment. We can combine the steps (3.1.14) and (3.1.15) into a single step, in which we use the following formula:

$$^{\alpha}\Delta = \sum_{t=1}^{T} f_t(\alpha) \cdot {}^{0}\Delta_t. \qquad (3.1.16)$$

This new algorithm is much faster. The original algorithm computed the formula (1.8.4) eleven times. The new algorithm uses the corresponding formula (3.1.13) (the analogue of the formula (1.8.4)) T times, i.e., as many times as there are types. All the other computations are much faster, since they do not grow with the input size n.

Thus, if the number T of different types is smaller than eleven, the new methods is much faster.

For example, if we have $T = 2$ different types, e.g., intervals and triangular membership functions, then we get a $\dfrac{11}{2} = 5.5$ times speedup.

Conclusion. We can therefore conclude that sometimes, it is beneficial to process different types of uncertainty separately – namely, it is beneficial when we have ten or fewer different types of uncertainty. The fewer types of uncertainty we have, the faster the resulting algorithm.

3.2   Fuzzy Data Processing Beyond min t-Norm

Usual algorithms for fuzzy data processing – based on the usual form of Zadeh's extension principle – implicitly assume that we use the min "and"-operation (t-norm). It is known, however, that in many practical situations, other t-norms more adequately describe human reasoning. It is therefore desirable to extend the usual algorithms to situations when we use t-norms different from min. Such an extension is provided in this section.

Results presented in this section first appeared in [81].

3.2.1 Need for Fuzzy Data Processing

Need for data processing. In many real-life situations, we are interested in the value of a quantity y which are difficult (or even impossible) to measure directly. For example, we may be interested in the distance to a faraway star, in the amount of oil in a given well, or in tomorrow's weather.

Since we cannot measure the quantity y directly, the way to estimate this value is to measure related easier-to-measure quantities $x_1, \ldots, x_n$, and then to use the known relation $y = f(x_1, \ldots, x_n)$ between the desired quantity y and the quantities x_i to estimate y; see, e.g, [82].

For example, since we cannot directly measure the distance to a faraway star, we can measure the directions x_i to this star in two different seasons, when the Earth is on the opposite sides of its Solar orbit, and then use trigonometry to find the desired distance. Since we cannot directly measure the amount y of oil in the well, we can measure the results of artificially set seismic waves propagating through the corresponding region, and then use the known equations of seismic wave propagation to compute y. Since we cannot directly measure tomorrow's temperature y, to estimate y, we measure temperature, moisture, wind speed, and other meteorological characteristics in the vicinity of our area, and use the known equations of aerodynamics and thermal physics to predict tomorrow's weather.

The computation of $y = f(x_1, \ldots, x_n)$ based on the results of measuring x_i is known as *data processing*.

In some cases, the data processing algorithm $f(x_1, \ldots, x_n)$ consists of direct application of a formula, but in most cases, we have a rather complex algorithm – such as solving a system of partial differential equations.

Need for fuzzy data processing. In some cases, instead of the measurement results $x_1, \ldots, x_n$, we have expert estimates for the corresponding quantities. Experts can rarely describe their estimates by exact numbers, they usually describe their estimates by using imprecise ("fuzzy") words from natural language. For example, an expert can say that the temperature is *warm*, or that the temperature is *around 20*.

To process such imprecise natural-language estimates, L. Zadeh designed the technique of *fuzzy logic*; see, e.g., [28, 61, 108]. In this technique, each imprecise word like "warm" is described by assigning, to each possible value of the temperature x, the degree to which the expert believes that this particular temperature is warm.

This degree can be obtained, e.g., by asking an expert to mark, on a scale from 0 to 10, where 0 means no warm and 10 means warn, how much x degrees means warm. Then, if for some temperature x, the expert marks, say, 7 on a scale from 0 to 10, we assign the degree $\mu(x) = 7/10$.

The function that assigns, to each possible value of a quantity x, the degree $\mu(x)$ to which the quantity satisfies the expert's estimate, is known as the *membership function*.

If for each of the inputs $x_1, \ldots, x_n$, we only have expert estimates, then what can we say about $y = f(x_1, \ldots, x_n)$? Computing appropriate estimates for y is known as *fuzzy data processing*.

To perform fuzzy data processing, we need to deal with propositional connectives such as "and" and "or". In fuzzy data processing, we know that $y = f(x_1, \ldots, x_n)$, and we have fuzzy information about each of the inputs x_i. Based on this information, we need to estimate the corresponding degrees of confidence in different values y.

A value y is possible if for some tuple $(x_1, \ldots, x_n)$ for which $y = f(x_1, \ldots, x_n)$, x_1 is a possible value of the first variable, *and* x_2 is a possible value of the second variable, $\ldots$, *and* x_n is a possible value of the nth variable. So, to describe the degree to which each value y is possible, we need to be able to describe our degrees of confidence in statements containing propositional connectives such as "and" and "or".

Dealing with "and"- and "or" in fuzzy logic is not as easy as in the 2-valued case. In the traditional 2-valued logic, when each statement is either true or false, dealing with "and" and "or" connectives is straightforward: if we know the truth values of two statements A and B, then these truth values uniquely determine the truth values of the propositional combinations $A \& B$ and $A \vee B$.

In contrast, when we deal with degrees of confidence, the situation is not as straightforward. For example, for a fair coin, we have no reason to be more confident that it will fall head or tail. Thus, it makes sense to assume that the expert's degrees of confidence $a = d(A)$ and $b = d(B)$ in statement $A =$"coin falls head" and $B =$"coin falls tail" are the same: $a = b$.

Since a coin cannot at the same time fall head and tail, the expert's degree of belief that both A and B happen is clearly 0: $d(A \& B) = 0$. On the other hand, if we take $A' = B' = A$, then clearly $A' \& B'$ is equivalent to A and thus, $d(A' \& B') = d(A) > 0$. This is a simple example where for two different pairs of statements, we have the same values of $d(A)$ and $d(B)$ but different values of $d(A \& B)$:

- in the first case, $d(A) = d(B) = a$ and $A(A \& B) = 0$, while
- in the second case, $d(A') = d(B') = a$, but $d(A' \& B') = a > 0$.

This simple example shows that, in general, the expert's degree of confidence in a propositional combination like $A \& B$ or $A \vee B$ is not uniquely determined by his/her degrees of confidence in the statements A and B. Thus, ideally, in addition to eliciting, from the experts, the degrees of confidence in all basic statements $A_1, \ldots, A_n$, we should also elicit from them degrees of confidence in all possible propositional combinations. For example, for each subset

$$I \subseteq \{1, \ldots, n\},$$

we should elicit, from the expert, his/her degree of confidence in a statement $\&_{i \in I} A_i$.

The problem is that there are 2^n such statement (and even more if we consider different propositional combinations). Even of a small size knowledge base, with $n = 30$ statements, we thus need to ask the expert about $2^{30} \approx 10^9$ different propositional combinations – this is clearly not practically possible.

Need for "and"- and "or"-operations. Since we cannot elicit the degrees of confidence for all propositional combinations from an expert, it is therefore desirable to estimate the degree of confidence in a propositional combination like $A \& B$ or $A \vee B$ from the expert's degrees of confidence $a = d(A)$ and $b = d(B)$ in the original statements A and B – knowing very well that these are *estimates*, not exactly the original expert's degrees of confidence in the composite statements.

The corresponding estimate for $d(A \& B)$ will be denoted by $f_{\&}(a, b)$. It is known as an *"and"-operation*, or, alternatively, as a *t-norm*. Similarly, the estimate for $d(A \vee B)$ will be denoted by $f_{\vee}(a, b)$. This estimate is called an *"or"-operation*, or a *t-conorm*.

There are reasonable conditions that these operations should satisfy. For example, since "A and B" means the same as "B and A", it makes sense to require that the estimates for $A \& B$ and $B \& A$ are the same, i.e., that $f_{\&}(a, b) = f_{\&}(b, a)$ for all a and b. Similarly, we should have $f_{\vee}(a, b) = f_{\vee}(b, a)$ for all a and b. The fact that $A \& (B \& C)$ means the same as $(A \& B) \& C$ implies that the corresponding estimates should be the same, i.e., that we should have $f_{\&}(a, f_{\&}(b, c)) = f_{\&}(f_{\&}(a, b), c)$ for all a, b, and c. There are also reasonable monotonicity requirements.

There are many different "and"- and "or"-operations that satisfy all these requirements. For example, for "and", we have the min-operation $f_{\&}(a, b) = \min(a, b)$. We also have an *algebraic product* operation $f_{\&}(a, b) = a \cdot b$. In addition, we can consider general *Archimedean operations*

$$f_{\&}(a, b) = f^{-1}(f(a) \cdot f(b)),$$

for some strictly increasing continuous function $f(x)$.

Similarly, for "or", we have the max-operation $f_{\vee}(a, b) = \max(a, b)$, we have *algebraic sum*

$$f_{\vee}(a, b) = a + b - a \cdot b = 1 - (1 - a) \cdot (1 - b),$$

and we have general Archimedean operations

$$f_{\vee}(a, b) = f^{-1}(f(a) + f(b) - f(a) \cdot f(b)).$$

Which "and" and "or"-operations should we select? Our objective is to describe the expert's knowledge. So, it is reasonable to select "and"- and "or"-operations which most adequately describe the reasoning of this particular expert (or this particular group of experts).

Such a selection was first made by Stanford researchers who designed the world's first expert system MYCIN – for curing rare blood diseases; see, e.g., [8]. It is interesting to mention that when the researchers found the "and"- and "or"-operations that best describe the reasoning of medical experts, they thought that they have found general laws of human reasoning. However, when they tried the same operations on another area of knowledge – geophysics – it turned out that different "and'- and "or'-operations are needed.

In hindsight, this difference make perfect sense: in medicine, one has to be very cautious, to avoid harming a patient. You do not start a surgery unless you are absolutely sure that the diagnosis is right. If a doctor is not absolutely sure, he or she recommends additional tests. In contrast, in geophysics, if you have a reasonable degree of confidence that there is oil in a given location, you dig a well – and you do not wait for a 100% certainty: a failed well is an acceptable risk.

What this experience taught us is that in different situations, different "and"- and "or"-operations are more adequate.

Let us apply "and"- and "or"-operation to fuzzy data processing. Let us go back to fuzzy data processing. To find the degree to which y is a possible value, we need to consider all possible tuples $(x_1, \ldots, x_n)$ for which $y = f(x_1, \ldots, x_n)$. The value y is possible if one of these tuples is possible, i.e., either the first tuple is possible *or* the second tuple is possible, etc.

Once we know the degree $d(x)$ to which each tuple $x = (x_1, \ldots, x_n)$ is possible, the degree to which y is possible can then be obtained by applying "or"-operation $f_\vee(a, b)$ to all these degrees:

$$d(y) = f_\vee \{d(x_1, \ldots, x_n) : f(x_1, \ldots, x_n)\}.$$

Usually, there are infinitely many such tuples x for which $f(x) = y$. For most "or"-operations, e.g., for

$$f_\&(a, b) = a + b - a \cdot b = 1 - (1 - a) \cdot (1 - b),$$

if we combine infinitely many terms, we get the same meaningless value 1. In effect, the only widely used operation for which this is not happening is $f_\vee(a, b) = \max(a, b)$. Thus, it is makes sense to require that

$$d(y) = \max\{d(x_1, \ldots, x_n) : f(x_1, \ldots, x_n) = y\}.$$

To fully describe this degree, we thus need to describe the degree $d(x_1, \ldots, x_n)$ to which a tuple $x = (x_1, \ldots, x_n)$ is possible. A tuple $(x_1, \ldots, x_n)$ is possible if x_1 is a possible values of the first variable, *and* x_2 is a possible of the second variable, etc.

We know the expert's degree of confidence that x_1 is a possible value of the first variable: this described by the corresponding membership function $\mu_1(x_1)$. Similarly, the expert's degree of confidence that x_2 is a possible value of the second variable is equal to $\mu_2(x_2)$, etc. Thus, to get the degree to which the tuple $(x_1, \ldots, x_n)$ is possible, we need to use an appropriate "and"-operation $f_\&(a, b)$ to combine these degrees $\mu_i(x_i)$:

$$d(x_1, \ldots, x_n) = f_\&(\mu_1(x_1), \ldots, \mu_n(x_n)).$$

Substituting this expression into the above formula for $\mu(y) = d(y)$, we arrive at the following formula.

General form of Zadeh's extension principle. If we know the the quantity y is related to the quantities $x_1, \ldots, x_n$ by a relation $y = f(x_1, \ldots, x_n)$, and we know the membership functions $\mu_i(x_i)$ that describe each inputs x_i, then the resulting membership function for y takes the form

$$\mu(y) = \max\{f_\&(\mu_1(x_1), \ldots, \mu_n(x_n)) : f(x_1, \ldots, x_n) = y\}.$$

This formula was originally proposed by Zadeh himself and is thus known as *Zadeh's extension principle*.

Algorithms for fuzzy data processing: what is already known and what we do in this section. Usually, Zadeh's extension principle is applied to the situations in which we use the min "and"-operation $f_\&(a, b) = \min(a, b)$. In this case, Zadeh's extension principle takes a simplified form

$$\mu(y) = \max\{\min(\mu_1(x_1), \ldots, \mu_n(x_n)) : f(x_1, \ldots, x_n) = y\}.$$

For this case, there are efficient algorithms for computing $\mu(y)$. (We will mention these algorithms in the nest section.)

The problem is that, as it is well known, in many practical situations, other "and"-operations more adequately describe expert reasoning [28, 61]. It is therefore desirable to generalize the existing efficient algorithms for fuzzy data processing from the case of min to the general case of an arbitrary "and"-operation. This is what we do in this section.

3.2.2 Possibility of Linearization

In most practical cases, measurement and estimation errors are relatively small. Before we get deeper into algorithms, let us notice that in the above text, we consider the generic data processing functions $f(x_1, \ldots, x_n)$.

In the general case, this is what we have to do. However, in most practical situations, the expert uncertainty is relatively small – just like measurement errors are usually relatively small, small in the sense that the squares of the measurement or estimation errors are much smaller than the errors themselves; see, e.g., [82].

Indeed, if we have a measurement or an expert estimate with accuracy 10% (i.e., 0.1), then the square of this inaccuracy is 0.01, which is much smaller than the original measurement or estimation error. If we measure or estimate with an even higher accuracy, e.g., 5% or 1%, then the quadratic term is even much more small than the measurement or estimation error.

Possibility of linearization. In such situations, instead of considering general functions $f(x_1, \ldots, x_n)$, we can expand the corresponding data processing function in Taylor series around the estimates $\tilde{x}_1, \ldots, \tilde{x}_n$, and keep only linear terms in this expansion – thus, ignoring the quadratic terms. As a result, the original expression

$$f(x_1, \ldots, x_n) = f(\tilde{x}_1 + \Delta x_1, \ldots, \tilde{x}_n + \Delta x_n),$$

where we denoted $\Delta x_i \stackrel{\text{def}}{=} x_i - \tilde{x}_i$, is replace by its linear approximation:

$$f(x_1, \ldots, x_n) = f(\tilde{x}_1 + \Delta x_1, \ldots, \tilde{x}_n + \Delta x_n) \approx c_0 + \sum_{i=1}^{n} c_i \cdot \Delta x_i,$$

where $c_0 \stackrel{\text{def}}{=} f(\tilde{x}_1, \ldots, \tilde{x}_n)$ and $c_i \stackrel{\text{def}}{=} \dfrac{\partial f}{\partial x_i}_{|x_i = \tilde{x}_i}$. Substituting $\Delta x_i = x_i - \tilde{x}_i$ into the above formula, we get a linear dependence

$$f(x_1, \ldots, x_n) = a_0 + \sum_{i=1}^{n} c_i \cdot x_i,$$

where $a_0 \stackrel{\text{def}}{=} c_0 - \sum_{i=1}^{n} c_i \cdot \tilde{x}_i$.

Zadeh's extension principle in case of linearization. For linear functions, the general Zadeh's extension principle takes the form

$$\mu(y) = \max \left\{ f_\&(\mu_1(x_1), \ldots, \mu_n(x_n)) : a_0 + \sum_{i=1}^{n} c_i \cdot x_i = y \right\}.$$

In particular, the min-based extension principle takes the form

$$\mu(y) = \max \left\{ \min(\mu_1(x_1), \ldots, \mu_n(x_n)) : a_0 + \sum_{i=1}^{n} c_i \cdot x_i = y \right\}.$$

In the linearized case, we can reduce a general problem to several sums of two variables. From the computational viewpoint, linearization has an additional advantage: it enables us to reduce a general problem, with many inputs, to a sequence of problems with only two inputs. To be more precise, we can reduce it to the problem of computing the sum of two variables.

Indeed, our goal is to compute the membership degrees corresponding to the sum

$$y = a_0 + c_1 \cdot x_1 + c_2 \cdot x_2 + \cdots + c_n \cdot x_n.$$

To simplify this problem, let us recall how this expression would be computed on a computer.

- First, we will compute $y_1 \stackrel{\text{def}}{=} c_1 \cdot x_1$, then the first intermediate sum $s_1 = a_0 + y_1$.

- After that, we will compute the second product $y_2 = c_2 \cdot x_2$, and the second intermediate sum $s_2 = s_1 + y_2$.
- Then, we compute the third product $y_3 = c_3 \cdot x_3$, and the third intermediate sum $s_3 = s_2 + y_3$.
- ...
- At the end, we compute $y_n = c_n \cdot x_n$ and $y = s_{n-1} + y_n$.

Let us follow the same pattern when computing the membership function $\mu(y)$. First, let us find the membership functions $\nu_i(y_i)$ corresponding to $y_i = c_i \cdot x_i$. For each y_i, there is only one value x_i for which $y_i = c_i \cdot x_i$ – namely, the value $x_i = y_i/c_i$ – so in Zadeh's formula, there is no need to apply "or" or "and", we just have

$$\nu_i(y_i) = \mu_i(y_i/c_i).$$

Similarly, for $s_1 = a_0 + y_1$, we have $\eta_1(s_1) = \nu_1(s_1 - a_0) = \mu_1((s_1 - a_0)/c_1)$. Now, we can do the following:

- first, based on the membership functions $\eta_1(s_1)$ and $\nu_2(y_2)$, we find the membership function $\eta_2(s_2)$ corresponding to the sum $s_2 = s_1 + y_2$,
- then, based on the membership functions $\eta_2(s_2)$ and $\nu_3(y_3)$, we find the membership function $\eta_3(s_3)$ corresponding to the sum $s_3 = s_2 + y_3$,
- etc.
- until, based on the membership functions $\eta_{n-1}(s_{n-1})$ and $\nu_n(y_n)$, we find the desired membership function $\mu(y)$ corresponding to the sum $y = s_{n-1} + y_n$.

On each step, we need to several times apply Zadeh's extension principle to the sum of two variables.

For the sum of two variables, these formulas have the following simplified form.

Zadeh's extension principle: case of the sum of two variables. For a general "and"-operation $f_\&(a, b)$, when we know the membership functions $\mu_1(x_1)$ and $\mu_2(x_2)$ describing two quantities x_1 and x_2, then, once we have selected x_1, the value x_2 for which $x_1 + x_2 = y$ can be easily described as $x_2 = y - x_1$. Thus, the membership function $\mu(y)$ for the sum $y = x_1 + x_2$ takes the form

$$\mu(y) = \max_{x_1} f_\&(\mu_1(x_1), \mu_2(y - x_1)).$$

In particular, for $f_\&(a, b) = \min(a, b)$, we have

$$\mu(y) = \max_{x_1} \min(\mu_1(x_1), \mu_2(y - x_1)).$$

In the following text, this is what we will consider: algorithms for computing these two formulas.

3.2.3 Efficient Fuzzy Data Processing for the **min** *t-Norm: Reminder*

Our objective is to come up with efficient algorithm for fuzzy data processing for a general "and"-operation.

To come up with such an algorithm, let us recall how fuzzy data processing is performed in the case of the min t-norm.

What is the input for the algorithm. To describe each of the two input membership functions $\mu_1(x_1)$ and $\mu_2(x_2)$, we need to describe their values at certain number of points $v_0 < v_1 < \cdots < v_{N-1}$, where N is the total number of these points. Usually, these points are equally spaced, i.e., $v_i = v_0 + i \cdot \Delta$ for some $\Delta > 0$. Thus, the inputs consist of $2N$ values $\mu_i(v_j)$ corresponding to $i = 1, 2$ and to $j = 0, 1, \ldots, N - 1$.

First – naive – algorithm for fuzzy data processing. When the values of each of the two variables x_1 and x_2 go from v_0 to $v_{N-1} = v_0 + (N - 1) \cdot \Delta$, their sum $y = x_1 + x_2$ takes values $w_0 \stackrel{\text{def}}{=} 2v_0$, $w_1 = w_0 + \Delta = 2v_0 + \Delta, \ldots$, all the way to $w_{2(N-1)} = 2v_{N-1} = 2v_0 + 2 \cdot (N - 1) \cdot \Delta$.

Each value w_k can be represented as the sum $v_i + v_{k-i}$ of possible values of x_1 and x_2 when $0 \leq i \leq N - 1$ and $0 \leq k - i \leq N - 1$, i.e., when

$$\max(k + 1 - N, 0) \leq i \leq \min(k, N - 1).$$

Thus, we have

$$\mu(w_k) = \max(\min(\mu_1(v_i), \mu_2(v_{k-i})) : \max(k + 1 - N, 0) \leq i \leq \min(k, N - 1)\}.$$

What is the computational complexity of this naive algorithm. The computational complexity of an algorithm is usually gauged by its number of computational steps – which is roughly proportional to the overall computation time.

In the above case, for each k, we have up to N different values i:

- exactly N for $k = N - 1$,
- $N - 1$ for $k = N - 2$ and $k = N$,
- $N - 2$ values i for $k = N - 3$ and $k = N + 1$,
- etc.

For each k and i, we need one min operation, to compute the minimum

$$\min(\mu_1(v_i), \mu_2(v_{k-i})).$$

Then, we need to find the maximum of all these numbers. So, we need N steps to compute the value w_{N-1}, $N-1$ steps to compute the values w_{N-2} and w_N, etc. Overall, we therefore need

$$N = 2 \cdot (N-1) + 2 \cdot (N-2) + \cdots + 2 \cdot 1 = N + 2 \cdot ((N-1) + (N-2) + \cdots + 1) =$$

$$N + 2 \cdot \frac{N \cdot (N-1)}{2} = N + N \cdot (N-1) = N^2$$

computational steps.

It is possible to compute $\mu(y)$ faster, by using α-cuts. It is known that for the case of the min t-norm, it is possible to compute $\mu(y)$ faster. This possibility comes from considering α-*cuts*

$$\mathbf{x}_i(\alpha) \stackrel{\text{def}}{=} \{x_i : \mu_i(x_i) \geq \alpha\}.$$

Indeed, according to Zadeh's formula, $\mu(y) \geq \alpha$ if and only if there exists values x_1 and x_2 for which $x_1 + x_2 = y$ and $\min(\mu_1(x_1), \mu_2(x_2)) \geq \alpha$, i.e., equivalently, for which $\mu_1(x_1) \geq \alpha$ and $\mu_2(x_2) \geq \alpha$.

Thus, the α-cut for y is equal to the set of all possible values $y = x_1 + x_2$ when x_1 belongs to the α-cut for x_1 and x_2 belongs to the α-cut for x_2:

$$\mathbf{y}(\alpha) = \{x_1 + x_2 : x_1 \in \mathbf{x}_1(\alpha) \,\&\, x_2 \in \mathbf{x}_2(\alpha)\}.$$

In many practical situations, each of the membership functions $\mu_i(x_i)$ increases up to a certain value, then decreases. In such situations, each α-cut is an interval: $\mathbf{x}_i(\alpha) = [\underline{x}_i(\alpha), \overline{x}_i(\alpha)]$. If we know that x_1 belongs to the interval $[\underline{x}_1(\alpha), \overline{x}_1(\alpha)]$ and that x_2 belongs to the interval $[\underline{x}_2(\alpha), \overline{x}_2(\alpha)]$, then possible values of $y = x_1 + x_2$ form the interval

$$[\underline{y}(\alpha), \overline{y}(\alpha)] = [\underline{x}_1(\alpha) + \underline{x}_2(\alpha), \overline{x}_1(\alpha) + \overline{x}_2(\alpha)].$$

(It is worth mentioning that this formula is a particular case of *interval arithmetic*; see, e.g., [26, 58].)

Thus, instead of N^2 elementary operations, we only need to perform twice as many operations as there are possible levels α. When we use $2N$ values $\mu_i(x_i)$, we can have no more than $2N$ different values α, so the computation time is

$$O(N) \ll N^2.$$

This is the reason why α-cuts – and interval computations – are mostly used in fuzzy data processing.

3.2.4 Efficient Fuzzy Data Processing Beyond min t-Norm: the Main Result of This Section

Now that we recalled how fuzzy data processing can be efficiently computed for the min t-norm, let us find out how we can efficiently compute it for the case of a general t-norm. Reminder: we know the membership functions $\mu_1(x_1)$ and $\mu_2(x_2)$, we want to estimate the values

$$\mu(y) = \max_{x_1} f_\&(\mu_1(x_1), \mu_2(y - x_1)).$$

Naive algorithm. Similar to the case of the min t-norm, the naive (straightforward) application of this formula leads to an algorithm that requires N^2 steps: the only difference is that now, we use $f_\&$ instead of min.

Let us show that for t-norms different from min, it is also possible to compute $\mu(y)$ faster.

Reducing to the case of a product t-norm. It is know that Archimedean t-norms, i.e., t-norms of the type $f^{-1}(f(a) \cdot f(b))$ for monotonic $f(x)$, are *universal approximators*, in the sense that for every t-norm and for every $\varepsilon > 0$, there exists an Archimedean t-norm whose value is ε-close to the given one for all possible a and b; see, e.g., [60].

Thus, from the practical viewpoint, we can safely assume that the given t-norm is Archimedean. For the Archimedean t-norm, the above formula takes the form

$$\mu(y) = \max_{x_1} f^{-1}(f(\mu_1(x_1)) \cdot f(\mu_2(y - x_1))).$$

The inverse function f^{-1} is monotonic, thus, the largest values of $f^{-1}(z)$ is attained when z is the largest. So, we have

$$\mu(y) = f^{-1}(v(y)),$$

where we denoted

$$v(y) = \max_{x_1} f(\mu_1(x_1)) \cdot f(\mu_2(y - x_1)).$$

So, if we denote $v_1(x_1) \stackrel{\text{def}}{=} f(\mu_1(x_1))$ and $v_2(x_2) \stackrel{\text{def}}{=} f(\mu_2(x_2))$, we arrive at the following problem: given functions $v_1(x_1)$ and $v_2(x_2)$, compute the function

$$v(y) = \max_{x_1}(v_1(x_1) \cdot v_2(y - x_1)).$$

This is exactly the original problem for the product t-norm. Thus, fuzzy data processing problem for a general t-norm can indeed be reduced to the problem of fuzzy data processing for the product t-norm.

Let us simplify our problem. Multiplication can be simplified if we take logarithm of both sides; then it is reduced to addition. This is why logarithms were invented in the first place – and this is why they were successfully used in the slide rule, the main computational tool for the 19 century and for first half of the 20 century – to make multiplication faster.

The values $v_i(x_i)$ are smaller than 1, so their logarithms are negative. To make formulas easier, let us use positive numbers, i.e., let us consider the values $\ell_1(x_1) \stackrel{\text{def}}{=} -\ln(v_1(x_1))$, $\ell_2(x_2) \stackrel{\text{def}}{=} -\ln(v_2(x_2))$, and $\ell(y) \stackrel{\text{def}}{=} -\ln(v(y))$. Since we changed signs, max becomes min, so we get the following formula:

$$\ell(y) = \min_{x_1}(\ell_1(x_1) + \ell_2(y - x_1)).$$

We have thus reduced our problem to a known problem in convex analysis. The above formula is well-known in *convex analysis* [87]. It is known as the *infimal convolution*, or an *epi-sum*, and usually denoted by

$$\ell = \ell_1 \,\square\, \ell_2.$$

There are efficient algorithms for solving this convex analysis problem. At least for situations when the functions $\ell_i(x_i)$ are convex (and continuous), there is an efficient algorithm for computing the infimal convolution. This algorithm is based on the use of *Legendre–Fenchel transform*

$$\ell^*(s) = \sup_x(s \cdot x - \ell(x)).$$

Specifically, it turns out that the Legendre transform of the infimal convolution is equal to the sum of the Legendre transforms:

$$(\ell_1 \,\square\, \ell_2)^* = \ell_1^* + \ell_2^*,$$

and that to reconstruct a function form its Legendre transform, it is sufficient to apply the Legendre transform once again:

$$\ell = (\ell^*)^*.$$

Thus, we can compute the infimal convolution as follows:

$$\ell_1 \,\square\, \ell_2 = (\ell_1^* + \ell_2^*)^*.$$

Similarly, we can compute the infimal composition of several functions

$$\ell_1 \,\square\, \cdots \,\square\, \ell_n = \min\{\ell_1(y_1) + \cdots + \ell_n(y_n) : y_1 + \cdots + y_n = y\},$$

as

$$\ell_1 \,\square\, \cdots \,\square\, \ell_n = (\ell_1^* + \cdots + \ell_n^*)^*.$$

There exists a fast (linear-time $O(N)$) algorithm for computing the Legendre transform; see, e.g., [47]. So, by using this algorithm, we can compute the results of fuzzy data processing in linear time.

Let us summarize the resulting algorithm.

3.2.5 Resulting Linear Time Algorithm for Fuzzy Data Processing Beyond min t-Norm

What is given and what we want to compute: reminder. We are given:

- a function $f(x_1, \ldots, x_n)$;
- n membership functions $\mu_1(x_1), \ldots, \mu_n(x_n)$; and
- an "and"-operation $f_\&(a, b)$.

We want to compute a new membership function

$$\mu(y) = \max\{f_\&(\mu_1(x_1), \ldots, \mu_n(x_n)) : f(x_1, \ldots, x_n) = y\}.$$

Algorithm.

- First, we represent the given "and"-operation in the Archimedean form $f_\&(a, b) = f^{-1}(f(a) \cdot f(b))$ for an appropriate monotonic function $f(x)$.

  In the following text, we will assume that we have algorithms for computing both the function $f(x)$ and the inverse function $f^{-1}(x)$.

- Then, for each i, we find the value $\tilde{x}_i$ for which $\mu_i(x_i)$ attains its largest possible value.

  For normalized membership functions, this value is $\mu_i(\tilde{x}_i) = 1$.

- We then compute the values $c_0 = f(\tilde{x}_1, \ldots, \tilde{x}_n)$ and $c_i = \dfrac{\partial f}{\partial x_i}\Big|_{x_i = \tilde{x}_i}$, and

$$a_0 = c_0 - \sum_{i=1}^{n} c_i \cdot \tilde{x}_i.$$

In the following text, we will then use a linear approximation

$$f(x_1, \ldots, x_n) = a_0 + \sum_{i=1}^{n} c_i \cdot x_i.$$

- After that, we compute the membership functions

$$\nu_1(s_1) = u_1((s_1 - a_0)/c_1)$$

and $\nu_i(y_i) = \mu_i(y_i/c_i)$ for $i > 2$.

In terms of the variables $s_1 = a_0 + c_1 \cdot x_1$ and $y_i = c_i \cdot x_i$, the desired quantity y has the form $y = s_1 + y_2 + \cdots + y_n$.

- We compute the minus logarithms of the resulting functions:

$$\ell_i(y_i) = -\ln(\nu_i(y_i)).$$

- For each i, we then use the Fast Legendre Transform algorithm from [47] to compute ℓ_i^*.
- Then, we add all these Legendre transforms and apply the Fast Legendre Transform once again to compute the function

$$\ell = (\ell_1^* + \cdots + \ell_n^*)^*.$$

- This function $\ell(y)$ is equal to $\ell(y) = -\ln(\nu(y))$, so we can reconstruct $\nu(y)$ as

$$\nu(y) = \exp(-\ell(y)).$$

- Finally, we can compute the desired membership function $\mu(u)$ as

$$\mu(y) = f^{-1}(\nu(y)).$$

3.2.6 Conclusions

To process fuzzy data, we need to use Zadeh's extension principle. In principle, this principle can be used for any t-norm. However, usually, it is only used for the min t-norm, since only for this t-norm, an efficient (linear-time) algorithm for fuzzy data processing was known.

Restricting oneselves to min t-norm is not always a good idea, since it is known that in many practical situations, other t-norms are more adequate in describing expert's reasoning. In this section, we show that similar efficient linear-time algorithms can be designed for an arbitrary t-norm. Thus, it become possible to use a t-norm that most adequately describes expert reasoning in this particular domain – and still keep fuzzy data processing algorithm efficient.

3.3 How to Speed Up Processing of Probabilistic Data

In many real-life situations, uncertainty can be naturally described as a combination of several components, components which are described by probabilistic, fuzzy, interval, etc. granules. In such situations, to process this uncertainty, it is often beneficial to take this granularity into account by processing these granules separately and then combining the results.

In this section, on the example of probabilistic uncertainty, we show that granular computing can help even in situations when there is no such natural decomposition into granules: namely, we can often speed up processing of uncertainty if we first (artificially) decompose the original uncertainty into appropriate granules.

Results described in this section first appeared in [80].

3.3.1 Need to Speed Up Data Processing Under Uncertainty: Formulation of the Problem

Need for data processing. One of the main reasons for data processing is that we are interested in a quantity y which is difficult (or even impossible) to measure or estimate directly. For example, y can be a future value of a quantity of interest.

To estimate this value y, we:

- find easier-to-measure and/or or easier-to-estimate quantities $x_1, \ldots, x_n$ which are related to y by a known dependence $y = f(x_1, \ldots, x_n)$,
- measure or estimate x_i's, and
- use the known relation $y = f(x_1, \ldots, x_n)$ to predict y.

Need to take uncertainty into account. Due to measurement uncertainty, the measurement results $\tilde{x}_i$ are, in general, different from the actual values x_i of the corresponding quantities. This is even more true to the results of expert estimates.

Therefore, the value $\tilde{y} = f(\tilde{x}_1, \ldots, \tilde{x}_n)$ that we obtain by processing the measurement/estimation results is, in general, different from the desired value $y = f(x_1, \ldots, x_n)$. It is therefore important to estimate the resulting uncertainty $\Delta y \overset{\text{def}}{=} \tilde{y} - y$; see, e.g., [82].

Measurement or estimation errors are usually relatively small. Measurement and estimation errors are usually assumed to be relatively small, so that terms quadratic in measurement errors can be safely ignored [82].

If we expand the expression

$$\Delta y = f(\widetilde{x}_1, \ldots, \widetilde{x}_n) - f(\widetilde{x}_1 - \Delta x_1, \ldots, \widetilde{x}_n - \Delta x_n)$$

in Taylor series and ignore terms which are quadratic (or higher order) in terms of Δx_i, then we get

$$\Delta y = \sum_{i=1}^{n} c_i \cdot \Delta x_i,$$

where $c_i \overset{\text{def}}{=} \dfrac{\partial f}{\partial x_i}$.

This simplified expression enables us to estimate the uncertainty in the result y of data processing based on the known information about the uncertainties Δx_i.

How to describe uncertainty. For measurements, we usually have a large number of situations when we performed the measurement with our measuring instrument and we also measured the same quantity with some more accurate measuring instrument – so that we have a good record of past values of measurement errors. For example, we may record the temperature outside by a reasonably cheap not-very-accurate thermometer, and we can find the measurement errors by comparing these measurement results with accurate measurements performed at a nearby meteorological station.

Based on such a record, we can estimate the probability of different values of the measurement error. Thus, it is reasonable to assume that for each i, we know the distribution of the measurement error $\Delta x_i \overset{\text{def}}{=} \widetilde{x}_i - x_i$.

Measurement errors corresponding to different variables are usually independent. In this section, we consider an ideal case when we know:

- the exact dependence $y = f(x_1, \ldots, x_n)$,
- the probability distribution of each of the variables Δx_i, and
- the values c_i.

Thus, we know the probability distribution of each of the terms $t_i = c_i \cdot \Delta x_i$. So, we arrive at the following problem:

- we know the probability distributions of each of n independent random variables

$$t_1, \ldots, t_n,$$

- we are interested in the probability distribution of their sum $t = \sum_{i=1}^{n} t_i$.

How this problem is solved now. The usual way to represent a probability distribution is by its probability density function (pdf) $\rho(x)$. The pdf of the sum $t = t_1 + t_2$ of two independent random variables with pdfs $\rho_1(t_1)$ and $\rho_2(t_2)$ is equal to

$$\rho(t) = \int \rho_1(t_1) \cdot \rho_2(t - t_1) \, dt_1.$$

A straightforward way of computing each value $\rho(t)$ is by replacing the integral with a sum. If we use N different points, then we need N computations to compute the sum corresponding to each of the N points, thus we need the total of N^2 computation steps; see, e.g., [97].

A faster computation can be done if we use characteristic functions $\chi_i(\omega) \stackrel{\text{def}}{=} E[\exp(i \cdot \omega \cdot t_i)]$, where E denotes the expected value. Then, from $t = t_1 + t_2$, we conclude that

$$\exp(i \cdot \omega \cdot t) = \exp(i \cdot \omega \cdot t_1) \cdot \exp(i \cdot \omega \cdot t_2)$$

and thus, since t_1 and t_2 are independent, that

$$E[\exp(i \cdot \omega \cdot t)] = E[\exp(i \cdot \omega \cdot t_1)] \cdot E[\exp(i \cdot \omega \cdot t_2)],$$

i.e., $\chi(\omega) = \chi_1(\omega) \cdot \chi_2(\omega)$. Here:

- computing each characteristic function $\chi_i(\omega)$ by Fast Fourier Transform requires $O(N \cdot \ln(N))$ computational steps,
- computing point-by-point multiplication requires N steps, and
- the inverse Fourier Transform to reconstruct $\rho(t)$ from its characteristic function also takes $O(N \cdot \ln(N))$ steps.

So overall, we need $O(N \cdot \ln(N))$ steps, which is smaller than N^2.

Can we do it faster? For large N, the time N needed for point-wise multiplication is still huge, so it is reasonable to look for the ways to make it faster.

3.3.2 Analysis of the Problem and Our Idea

Processing can be faster if both distributions are normal. If both t_i are normally distributed, then we do not need to perform these computations: we know that the sum of two normal distributions with mean μ_i and variances V_i is also normal, with mean $\mu = \mu_1 + \mu_2$ and variance $V = V_1 + V_2$.

In this case, we need two computational steps instead of $O(N)$.

Other cases when we can speed up data processing. Same holds for any *infinitely divisible* distribution, with characteristic function

$$\chi(\omega) = \exp(i \cdot \mu \cdot \omega - A \cdot |\omega|^\alpha).$$

For example:

- For $\alpha = 2$, we get normal distribution.
- For $\alpha = 1$, we get Cauchy distribution, with the probability density function

$$\rho(x) = \frac{1}{\pi \cdot \Delta} \cdot \frac{1}{1 + \dfrac{(x - \mu)^2}{\Delta^2}}$$

for an appropriate $\Delta > 0$.

Indeed, in this case, once we know the distributions for t_1 and t_2, then, based on the corresponding characteristic functions $\chi_1(\omega) = \exp(\mathrm{i} \cdot \mu_1 \cdot \omega - A_1 \cdot |\omega|^\alpha)$ and $\chi_2(\omega) = \exp(\mathrm{i} \cdot \mu_2 \cdot \omega - A_2 \cdot |\omega|^\alpha)$, we can conclude that the characteristic function $\chi(\omega)$ for the sum $t_1 + t_2$ has the form

$$\chi(\omega) = \chi_1(\omega) \cdot \chi_2(\omega) = \exp(\mathrm{i} \cdot \mu \cdot \omega - A \cdot |\omega|^\alpha),$$

where $\mu = \mu_1 + \mu_2$ and $A = A_1 + A_2$. So, in this case too, we need two computational steps instead of $O(N)$:

- one step to add the means μ_i, and
- another step to add the values A_i.

Our idea. Our idea is to select several values $\alpha_1, \ldots, \alpha_k$ (e.g., $\alpha_1 = 1$ and $\alpha_2 = 2$) and approximate each random variable t_i by a sum

$$t_{a,i} = t_{i1} + \cdots + t_{ij} + \cdots + t_{ik}$$

of infinitely divisible random variables t_{ij} corresponding to the selected values of α_j.

The characteristic function $\chi_{ij}(\omega)$ for each variable t_{ij} has the form

$$\chi_{ij}(\omega) = \exp(\mathrm{i} \cdot \mu_{ij} \cdot \omega - A_{ij} \cdot |\omega|^{\alpha_j}).$$

Thus, the characteristic function $\chi_{a,i}(\omega)$ of the sum $t_{a,i} = \sum_{j=1}^{k} t_{ik}$ is equal to the product

$$\chi_{a,i}(\omega) = \prod_{j=1}^{k} \chi_{ij}(\omega) = \exp\left(\mathrm{i} \cdot \mu_i \cdot \omega - \sum_{j=1}^{k} A_{ij} \cdot |\omega|^{\alpha_j}\right),$$

where $\mu_i \overset{\text{def}}{=} \sum_{j=1}^{k} \mu_{ij}$.

From $\chi_1(\omega) \approx \chi_{a,1}(\omega)$ and $\chi_2(\omega) \approx \chi_{a,2}(\omega)$, we conclude that the characteristic function $\chi(\omega) = \chi_1(\omega) \cdot \chi_2(\omega)$ for the sum $t = t_1 + t_2$ is approximately equal to the product of the approximating characteristic functions:

$$\chi(\omega) \approx \chi_a(\omega) \overset{\text{def}}{=} \chi_{a,2}(\omega) \cdot \chi_{a,2}(\omega) = \exp\left(\mathrm{i} \cdot \mu \cdot \omega - \sum_{j=1}^{k} A_j \cdot |\omega|^{\alpha_j}\right),$$

where $\mu = \mu_1 + \mu_2$ and $A_j = A_{1j} + A_{2j}$.

In this case, to find the approximating distribution for the sum t, we need to perform $k + 1$ arithmetic operations instead of N:

- one addition to compute μ and
- k additions to compute k values $A_1, \ldots, A_k$.

Comment. A similar idea can be applied to the case of fuzzy uncertainty; see Sect. 3.1 for details.

3.3.3 How to Approximate

Natural idea: use Least Squares. We want to approximate the actual distribution $\rho_i(t)$ for each of the variables t_i by an approximate approximate distribution $\rho_{a,i}(t)$. A reasonable idea is to use the Least Squares approximation, i.e., to find a distribution $\rho_{a,i}(t)$ for which the value $\int (\rho_i(t) - \rho_{a,i}(t))^2 \, dt$ is the smallest possible.

Let us reformulate this idea in terms of the characteristic functions. The problem with the above idea is that while for $\alpha = 1$ and $\alpha = 2$, we have explicit expressions for the corresponding probability density function $\rho_{a,i}(t)$, we do not have such an expression for any other α. Instead, we have an explicit expression for the characteristic function $\chi(\omega)$. It is therefore desirable to reformulate the above idea in terms of characteristic functions.

We want to approximate the characteristic function $\chi_i(\omega)$ by an expression $\chi_{a,i}(\omega)$ of the type $\exp\left(-\sum_j c_j \cdot f_j(\omega)\right)$ for some fixed functions $f_j(\omega)$; in our case, $f_0(\omega) = -i \cdot \omega$ and $f_j(\omega) = |\omega|^{\alpha_j}$ for $j \geq 1$.

This can be done, since, due to Parceval theorem, the least squares (L^2) difference $\int (\rho_i(t) - \rho_{a,i}(t))^2 \, dt$ between the corresponding pdfs $\rho_i(t)$ and $\rho_{a,i}(t)$ is proportional to the least squares difference between the characteristic functions:

$$\int (\rho_i(t) - \rho_{a,i}(t))^2 \, dt = \frac{1}{2\pi} \cdot \int (\chi_i(\omega) - \chi_{a,i}(\omega))^2 \, d\omega.$$

So, minimizing the value $\int (\rho_i(t) - \rho_{a,i}(t))^2 \, dt$ is equivalent to minimizing

$$I \stackrel{\text{def}}{=} \int (\chi_i(\omega) - \chi_{a,i}(\omega))^2 \, d\omega.$$

How to approximate: computational challenge and its solution. The problem with the above formulation is that the Least Squares method is very efficient is we are looking for the coefficients of a linear dependence. However, in our case, the dependence of the expression $\chi_{a,i}(\omega)$ on the parameters μ_i and A_{ij} is non-linear, which makes computations complicated.

How can we simplify computations? We can borrow the idea from the case of normal distributions: in this case, we start with the maximum likelihood methods, in which we maximize the probability, and take negative logarithms of the pdfs – which results in the known Least Squares method [97]. In our more general case too, if we take the negative logarithm of the characteristic function, we get a linear function of the unknowns:

$$- \ln(\chi_{a,i}(\omega)) = -i \cdot \mu_i \cdot \omega + \sum_{j=1}^{k} A_{ij} \cdot |\omega|^{\alpha_j}.$$

To use this idea, let us reformulate the objective function

$$\int (\chi_i(\omega) - \chi_{a,i}(\omega))^2 \, d\omega$$

in terms of the difference between the negative logarithms. We are interested in situations in which the approximation is good, i.e., in which the difference $\varepsilon_i(\omega) \overset{\text{def}}{=} \chi_{a,i}(\omega) - \chi_i(\omega)$ is small. Then, $\chi_{a,i}(\omega) = \chi_i(\omega) + \varepsilon_i(\omega)$, hence

$$- \ln(\chi_{a,i}(\omega)) = - \ln(\chi_i(\omega) + \varepsilon_i(\omega)) = - \ln \left(\chi_i(\omega) \cdot \left(1 + \frac{\varepsilon_i(\omega)}{\chi_i(\omega)}\right) \right) =$$

$$- \ln(\chi_i(\omega)) - \ln \left(1 + \frac{\varepsilon_i(\omega)}{\chi_i(\omega)}\right).$$

Since $\varepsilon_i(\omega)$ is small, we can ignore terms which are quadratic and higher order in $\varepsilon_i(\omega)$ and get

$$\ln \left(1 + \frac{\varepsilon_i(\omega)}{\chi_i(\omega)}\right) \approx \frac{\varepsilon_i(\omega)}{\chi_i(\omega)}.$$

Thus, in this approximation,

$$(- \ln(\chi_i(\omega))) - (- \ln(\chi_{a,i}(\omega))) = \frac{\varepsilon_i(\omega)}{\chi_i(\omega)},$$

hence

$$\varepsilon_i(\omega) = \chi_{a,i}(\omega) - \chi_i(\omega) = \chi_i(\omega) \cdot ((- \ln(\chi_{a,i}(\omega))) - (- \ln(\chi_i(\omega)))),$$

so the minimized integral takes the form

$$I = \int (\chi_i(\omega) - \chi_{a,i}(\omega))^2 \, d\omega = \int \chi_i^2(\omega) \cdot ((- \ln(\chi_i(\omega))) - (- \ln(\chi_{a,i}(\omega)))^2 \, d\omega,$$

or, equivalently, the form

$$I = \int (f_i(\omega) - f_{a,i}(\omega))^2 \, d\omega,$$

where we denoted

$$f_i(\omega) \overset{\text{def}}{=} -\chi_i(\omega) \cdot \ln(\chi_i(\omega))$$

and

$$f_{a,i}(\omega) \overset{\text{def}}{=} -\chi_i(\omega) \cdot \ln(\chi_{a,i}(\omega)).$$

In our case

$$f_{a,i}(\omega) = -\mathrm{i} \cdot \mu_i \cdot \omega \cdot \chi_i(\omega) + \sum_{j=1}^{k} A_{ij} \cdot \chi_i(\omega) \cdot |\omega|^{\alpha_j}.$$

In other words, we need to find the coefficients c_k by applying the Least Squares method to the approximate equality

$$-\ln(\chi_i(\omega)) \cdot \chi_i(\omega) \approx -\mathrm{i} \cdot \mu_i \cdot \omega \cdot \chi_i(\omega) + \sum_{j=1}^{k} A_{ij} \cdot \chi_i(\omega) \cdot |\omega|^{\alpha_j}.$$

3.3.4 Resulting Algorithm

Problem: reminder. We know the probability distributions for $t_1 = c_1 \cdot \Delta x_1$ and $t_2 = c_2 \cdot \Delta x_2$, We want to find the probability distribution for

$$t = t_1 + t_2 = c_1 \cdot \Delta x_1 + c_2 \cdot \Delta x_2.$$

Motivations: reminder. By repeating this procedure several times, we get:

- the probability distribution for $c_1 \cdot \Delta x_1 + c_2 \cdot \Delta x_2 = \sum\limits_{i=1}^{2} c_i \cdot \Delta x_i$,
- then the distribution for $(c_1 \cdot \Delta x_1 + c_2 \cdot \Delta x_2) + c_3 \cdot \Delta x_2 = \sum\limits_{i=1}^{3} c_i \cdot \Delta x_i$,
- then the distribution for

$$\left(\sum_{i=1}^{3} c_i \cdot \Delta x_i \right) + c_4 \cdot \Delta x_4 = \sum_{i=1}^{4} c_i \cdot \Delta x_i,$$

etc.,

• until we get the desired probability distribution for the measurement error

$$\Delta y = \sum_{i=1}^{n} c_i \cdot \Delta x_i.$$

Preliminary step. We select the values $\alpha_1 < \cdots < \alpha_k$. For example, we can have these values uniformly distributed on the interval $[1, 2]$, by taking $\alpha_j = 1 + \dfrac{j - 1}{k - 1}$. For example:

• for $k = 2$, we get $\alpha_1 = 1$ and $\alpha_2 = 2$,
• for $k = 3$, we get $\alpha_1 = 1$, $\alpha_2 = 1.5$, and $\alpha_3 = 2$.

Comment. A (slightly) better selection of the values α_j is described in Sect. 3.2.6.

First step: computing characteristic functions. First, we apply Fast Fourier transform to the given probability distributions $\rho_i(t)$, and get the corresponding characteristic functions $\chi_i(\omega)$.

Second step: approximating characteristic functions. For each of the two characteristic functions, to find the parameters $\mu_i, A_{i1}, \ldots, A_{ik}$, we use the Least Squares method to solve the following system of approximate equations:

$$- \ln(\chi_i(\omega)) \cdot \chi_i(\omega) \approx \chi_{a,i}(\omega) \stackrel{\text{def}}{=} -i \cdot \mu_i \cdot \omega \cdot \chi_i(\omega) + \sum_{j=1}^{k} A_{ij} \cdot \chi_i(\omega) \cdot |\omega|^{\alpha_j}$$

for values $\omega = \omega_1, \omega_2, \ldots, \omega_N$.

Comment. If the resulting approximation error $\int (\chi_i(\omega) - \chi_{a,i}(\omega))^2 \, d\omega$ is too large, we can increase k – and thus, get a better approximation.

Final step: describing the desired probability distribution for $t = t_1 + t_2$. As a good approximation for the characteristic function $\chi(\omega)$ of the probability distribution for the sum $t = t_1 + t_2$, we can take the expression

$$\chi_a(\omega) = \exp\left(i \cdot \mu \cdot \omega - \sum_{j=1}^{k} A_j \cdot |\omega|^{\alpha_j} \right),$$

where $\mu = \mu_1 + \mu_2$ and $A_j = A_{1j} + A_{2j}$ for $j = 1, \ldots, k$.

3.3.5 Numerical Example

We tested our method on several examples, let us provide one such example.

Let us assume that t_1 is normally distributed with 0 mean and standard deviation 1. For this distribution, the characteristic function takes the form

$$\chi_1(\omega) = \exp\left(-\frac{1}{2} \cdot \omega^2\right).$$

As t_2, let us take the Laplace distribution, with probability density

$$\rho_2(t) = \frac{1}{2} \cdot \exp(-|t|)$$

and the characteristic function $\chi_2(\omega) = \dfrac{1}{1+\omega^2}$.

To approximate both distributions, we used $k = 3$, with $\alpha_1 = 1$, $\alpha_2 = 1.5$, and $\alpha_3 = 2$. In this case, the first distribution is represented exactly, with

$$\mu_1 = 0, \quad A_{11} = A_{12} = 0, \quad \text{and} \quad A_{13} = \frac{1}{2}.$$

To find the optimal approximation for the characteristic function of the Laplace distribution, we used the values ω uniformly distributed on the interval $[-5, 5]$. As a result, we get the following approximation:

$$\mu_2 = 0, \quad A_{21} = -0.162, \quad A_{22} = 1.237, \quad \text{and} \quad A_{23} = -0.398.$$

Thus, for the characteristic function of the sum $t = t_1 + t_2$ we get

$$\mu = 0, \quad A_1 = -0.162, \quad A_2 = 1.237, \quad \text{and} \quad A_3 = 0.102.$$

By applying the inverse Fourier transform to this distribution, we get an approximate probability density function $\rho_a(t)$ for the sum. The comparison between the actual probability distribution $\rho(t)$ and the approximate pdf $\rho_a(t)$ is given on Fig. 3.1. The corresponding mean square error $\sqrt{\int (\rho(t) - \rho_a(t))^2 \, dt}$ is equal to 0.01.

3.3.6 Non-uniform Distribution of α_j is Better

Idea. If we select two values α_j too close to each other, there will be too much correlation between them, so adding the function corresponding to the second value does not add much information to what we know from a function corresponding to the first value.

We are approximating a general function (logarithm of a characteristic function) as a linear combination of functions $|t|^{\alpha_j}$. If two values α_j and α_{j+1} are close, then

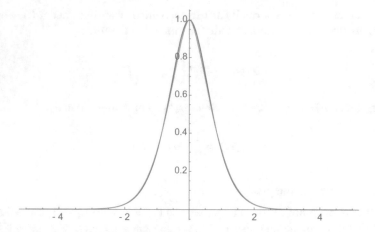

Fig. 3.1 How good is the proposed approximation

the function $|t|^{\alpha_{j+1}}$ can be well approximated by a term linear in $|t|^{\alpha_j}$, thus, the term proportional to $|t|^{\alpha_{j+1}}$ is not needed.

It therefore makes sense to select the values α_j in such as way that for each j, the part of $|t|^{\alpha_{j+1}}$ that cannot be approximated by terms proportional to $|t|^{\alpha_j}$ should be the largest possible.

Let us reformulate this idea in precise terms. For every two functions $f(t)$ and $g(t)$, the part of $g(t)$ which cannot be represented by terms $a \cdot f(t)$ (proportional to $f(t)$) can be described as follows. It is reasonable to describe the difference between the two functions $f(t)$ and $g(t)$ by the least squares (L^2) metric $\int (f(t) - g(t))^2 \, dt$. In these terms, the value of a function itself itself can be described as its distance from 0, i.e., as $\int (f(t))^2 \, dt$.

When we approximate a function $g(t)$ by a term $a \cdot f(t)$, then the remainder $g(t) - a \cdot f(t)$ has the value $\int (g(t) - a \cdot f(t))^2 \, dt$. The best approximation occurs when this value is the smallest, i.e., when it is equal to $\min_a \int (g(t) - a \cdot f(t))^2 \, dt$. Out of the original value $\int (g(t))^2 \, dt$, we have unrepresented the part equal to

$$\min_a \int (g(t) - a \cdot f(t))^2 \, dt.$$

Thus, the relative size of what cannot be represented by terms $a \cdot f(t)$ can be defined as a ratio

$$R(f(t), g(t)) = \frac{\min_a \int (g(t) - a \cdot f(t))^2 \, dt}{\int (g(t))^2 \, dt}.$$

Let us simplify the resulting expression. This expression can be simplified if we find the explicit expression for a for which the value $\int (g(t) - a \cdot f(t))^2 \, dt$ is the smallest possible. Differentiating the minimized expression with respect to a and equating the derivative to 0, we conclude that

$$- \int (g(t) - a \cdot f(t)) \cdot f(t) \, dt = 0,$$

i.e., that

$$a \cdot \int (f(t))^2 \, dt = \int f(t) \cdot g(t) \, dt,$$

and

$$a = \frac{\int f(t) \cdot g(t) \, dt}{\int (f(t))^2 \, dt}.$$

For this a, the value $\int (g(t) - a \cdot f(t))^2 \, dt$ takes the form

$$\int (g(t) - a \cdot f(t))^2 \, dt = \int (g(t))^2 \, dt - 2a \cdot \int f(t) \cdot g(t) \, dt + a^2 \cdot \int (f(t)) \, dt.$$

Substituting the above expression for a into this formula, we conclude that

$$\int (g(t) - a \cdot f(t))^2 \, dt = \int (g(t))^2 \, dt - \frac{2(\int f(t) \cdot g(t) \, dt)^2}{\int (f(t))^2 \, dt} + \frac{(\int f(t) \cdot g(t) \, dt)^2}{\int (f(t))^2 \, dt},$$

i.e., that

$$\int (g(t) - a \cdot f(t))^2 \, dt = \int (g(t))^2 \, dt - \frac{(\int f(t) \cdot g(t) \, dt)^2}{\int (f(t))^2 \, dt}.$$

Thus, the desired ratio takes the form

$$R(f(t), g(t)) \overset{\text{def}}{=} \frac{\min\limits_{a} \int (g(t) - a \cdot f(t))^2 \, dt}{\int (g(t))^2 \, dt} = 1 - \frac{(\int f(t) \cdot g(t) \, dt)^2}{(\int (f(t))^2 \, dt) \cdot (\int (g(t))^2 \, dt)}.$$

Thus, we arrive at the following optimization problem.

Resulting optimization problem. To make sure that the above remainders are as large as possible, it makes sense to find the values $\alpha_1^{\text{opt}} < \cdots < \alpha_k^{\text{opt}}$ that maximize the smallest of the remainders between the functions $f(t) = |t|^{\alpha_j}$ and $g(t) = |t|^{\alpha_{j+1}}$:

$$\min_j R\left(|t|^{\alpha_j^{\text{opt}}}, |t|^{\alpha_{j+1}^{\text{opt}}}\right) = \max_{\alpha_1 < \cdots < \alpha_k} \min_j R(|t|^{\alpha_j}, |t|^{\alpha_{j+1}}).$$

Solving the optimization problem. Let us consider an interval $[-T, T]$ for some T. Since the function is symmetric, it is sufficient to consider the values from $[0, T]$. For $f(t) = t^{\alpha}$ and $g(t) = t^{\beta}$, the integral in the numerator of the ratio is equal to

$$\int_0^T f(t) \cdot g(t)\, dt = \int_0^T t^\alpha \cdot t^\beta\, dt = \int_0^T t^{\alpha+\beta}\, dt = \frac{T^{\alpha+\beta+1}}{\alpha+\beta+1}.$$

Similarly, the integrals in the denominator take the form

$$\int_0^T f^2(t)\, dt = \int_0^T t^{2\alpha}\, dt = \frac{T^{2\alpha+1}}{2\alpha+1}$$

and

$$\int_0^T g^2(t)\, dt = \int_0^T t^{2\beta}\, dt = \frac{T^{2\beta+1}}{2\beta+1},$$

so

$$R = 1 - \frac{\dfrac{T^{2(\alpha+\beta+1)}}{(\alpha+\beta+1)^2}}{\dfrac{T^{2\alpha+1}}{2\alpha+1} \cdot \dfrac{T^{2\beta+1}}{2\beta+1}}.$$

One can see that the powers of T cancel each other, and we get

$$R = 1 - \frac{(2\alpha+1)\cdot(2\beta+1)}{(\alpha+\beta+1)^2},$$

or, equivalently, if we denote $r \stackrel{\text{def}}{=} \dfrac{\beta+0.5}{\alpha+0.5}$, we get

$$R = R(r) \stackrel{\text{def}}{=} 1 - 4 \cdot \frac{r}{(1+r)^2}.$$

The derivative of the function $R(r)$ is equal to

$$\frac{dR}{dr} = -4 \cdot \frac{(1+r)^2 - 2\cdot(1+r)}{(1+r)^4} = -4 \cdot \frac{(1+r)\cdot(1+r-2)}{(1+r)^4} =$$

$$4 \cdot \frac{(1+r)\cdot(r-1)}{(1+r)^4} = 4 \cdot \frac{r-1}{(1+r)^3}.$$

So this derivative is positive for all $r > 1$. Thus, the function $R(r)$ is monotonically increasing, and looking for the values α_j^{opt} for which $\min_j R(|t|^{\alpha_j}, |t|^{\alpha_{j+1}})$ is the largest is equivalent to looking for the values α_j^{opt} for which the smallest $\min_j \dfrac{\alpha_{j+1}+0.5}{\alpha_j+0.5}$ of the ratios $r = \dfrac{\alpha_{j+1}+0.5}{\alpha_j+0.5}$ attains the largest possible value:

$$\min_{j} \frac{\alpha_{j+1}^{opt} + 0.5}{\alpha_{j}^{opt} + 0.5} = \max_{\alpha_1 < \cdots < \alpha_k} \min_{j} \frac{\alpha_{j+1} + 0.5}{\alpha_j + 0.5}.$$

One can check that this happens when $\alpha_j + 0.5 = 1.5 \cdot \left(\frac{5}{3}\right)^{(j-1)/(k-1)}$. Indeed,

in this case, $\min_{j} \dfrac{\alpha_{j+1} + 0.5}{\alpha_j + 0.5} = \left(\dfrac{5}{3}\right)^{1/(k-1)}$. We cannot have it larger: if we had

$\min_{j} \dfrac{\alpha_{j+1} + 0.5}{\alpha_j + 0.5} > \left(\dfrac{5}{3}\right)^{k-1}$, then we would have $\dfrac{\alpha_{j+1} + 0.5}{\alpha_j + 0.5} > \left(\dfrac{5}{3}\right)^{k-1}$ for all j.
Here,

$$\alpha_k + 0.5 = (\alpha_1 + 0.5) \cdot \frac{\alpha_2 + 0.5}{\alpha_1 + 0.5} \cdot \frac{\alpha_3 + 0.5}{\alpha_2 + 0.5} \cdots \frac{\alpha_k + 0.5}{\alpha_{k-1} + 0.5}.$$

The first factor $\alpha_1 + 0.5$ is ≥ 1.5, each of the other $k - 1$ terms is greater than $\left(\dfrac{5}{3}\right)^{1/(k-1)}$, so for their product, we get

$$\alpha_k + 0.5 > 1.5 \cdot \left(\left(\frac{5}{3}\right)^{1/(k-1)}\right)^{k-1} = 1.5 \cdot \frac{5}{3} = 2.5,$$

while we assumed that all the values α_j are from the interval $[1, 2]$, and so, we should have $\alpha_k + 0.5 < 2.5$.

Resulting optimal values of α_j. Thus, the optimal way is to not to take the values uniformly distributed on the interval $[1, 2]$, but rather take the values

$$\alpha_j^{opt} = 1.5 \cdot \left(\frac{5}{3}\right)^{(j-1)/(k-1)} - 0.5$$

for which the logarithms $\ln(\alpha_j^{opt} + 0.5) = \dfrac{j-1}{k-1} \cdot \ln\left(\dfrac{5}{3}\right) = \ln(1.5)$ are uniformly distributed.

Comment. It is worth mentioning that there is intriguing connection between these values α_j and music: for example, the twelves notes on a usual Western octave correspond to the frequencies:

- f_1,
- $f_2 = f_1 \cdot 2^{1/12}$,
- $f_3 = f_1 \cdot 2^{2/12}, \ldots,$
- $f_{12} = f_1 \cdot 2^{11/12}$, and
- $f_{13} = f_1 \cdot 2$,

for which the logarithms $\ln(f_j)$ are uniformly distributed. Similar formulas exist for five-note and other octaves typical for some Oriental musical traditions.

3.4   Hypothetical Quantum-Related Negative Probabilities Can Speed Up Uncertainty Propagation Algorithms

One of the main features of quantum physics is that, as basic objects describing uncertainty, instead of (non-negative) probabilities and probability density functions, we have complex-valued probability amplitudes and wave functions. In particular, in quantum computing, negative amplitudes are actively used. In the current quantum theories, the actual probabilities are always non-negative. However, there have been some speculations about the possibility of actually negative probabilities. In this section, we show that such hypothetical negative probabilities can lead to a drastic speed up of uncertainty propagation algorithms.

Results described in this section first appeared in [74].

3.4.1   Introduction

From non-negative to more general description of uncertainty. In the traditional (non-quantum) physics, the main way to describe uncertainty – when we have several alternatives and we do not know which one is true – is by assigning probabilities p_i to different alternatives i.

The physical meaning of each probability p_i is that it represents the frequency with which the ith alternative appears in similar situations. As a result of this physical meaning, probabilities are always non-negative.

In the continuous case, when the number of alternatives is infinite, each possible alternative has 0 probability. However, we can talk:

- about probabilities of values being in a certain interval and, correspondingly,
- about the probability density $\rho(x)$ – probability per unit length or per unit volume.

The corresponding probability density function is a limit of the ratio of two non-negative values:

- probability and
- volume,

and is, thus, also always non-negative.

One of the main features of quantum physics is that in quantum physics, probabilities are no longer the basic objects for describing uncertainty; see, e.g., [18]. To describe a general uncertainty, we now need to describe the complex-valued *probability amplitudes* ψ_i corresponding to different alternatives i. In the continuous case:

- instead of a probability density function $\rho(x)$,
- we have a complex-valued *wave function* $\psi(x)$.

Non-positive and non-zero values of the probability amplitude and of the wave function are important: e.g., negative values of the amplitudes are actively used in many quantum computing algorithms; see, e.g., [63].

Can there be negative probabilities? In the current quantum theories, the actual probabilities are always non-negative. For example:

- the probability p_i of observing the ith alternative is equal to a non-negative number

$$p_i = |\psi_i|^2,$$

and
- the probability density function is equal to a non-negative expression

$$\rho(x) = |\psi(x)|^2.$$

However, there have been some speculations about the possibility of actually negative probabilities, speculations actively explored by Nobel-rank physicists such as Dirac and Feynman; see, e.g., [14, 17]. Because of the high caliber of these scientists, it makes sense to take these speculations very seriously.

What we do in this section. In this section, we show that such hypothetical negative probabilities can lead to a drastic speed up of uncertainty propagation algorithms.

3.4.2 Uncertainty Propagation: Reminder and Precise Formulation of the Problem

Need for data processing. In many practical situations, we are interested in the value of a physical quantity y which is difficult or even impossible to measure directly. For example, we may be interested:

- in tomorrow's temperature, or
- in a distance to a faraway star, or
- in the amount of oil in a given oil field.

Since we cannot measure the quantity y directly, a natural idea is:

- to measure easier-to-measure related quantities

$$x_1, \ldots, x_n,$$

and then
- to use the known relation

$$y = f(x_1, \ldots, x_n)$$

between these quantities to estimate y as

$$\tilde{y} = f(\tilde{x}_1, \ldots, \tilde{x}_n),$$

where $\tilde{x}_i$ denotes the result of measuring the quantity x_i.

For example:

- To predict tomorrow's temperature y:

  - we measure temperature, humidity, and wind velocity at different locations, and
  - we use the known partial differential equations describing atmosphere to estimate y.

- To measure a distance to a faraway star:

  - we measure the direction to this star in two different seasons, when the Earth is on different sides of the Sun, and then
  - we use trigonometry to find y based on the difference between the two measured directions.

In all these cases, the algorithm f transforming our measurement results into the desired estimate $\tilde{y}$ is an example of *data processing*.

Need for uncertainty propagation. Measurements are never absolutely accurate. The measurement result $\tilde{x}_i$ is, in general, somewhat different from the actual (unknown) value of the corresponding quantity x_i. As a result, even when the relation

$$y = f(x_1, \ldots, x_n)$$

is exact, the result $\tilde{y}$ of data processing is, in general, somewhat different from the the actual values $y = f(x_1, \ldots, x_n)$:

$$\tilde{y} = f(\tilde{x}_1, \ldots, \tilde{x}_n) \neq y = f(x_1, \ldots, x_n).$$

It is therefore necessary to estimate

- how accurate is our estimation $\tilde{y}$, i.e.,
- how big is the estimation error

$$\Delta y \overset{\text{def}}{=} \tilde{y} - y.$$

The value of Δy depends on how accurate were the original measurements, i.e., how large were the corresponding measurement errors

$$\Delta x_i \overset{\text{def}}{=} \tilde{x}_i - x_i.$$

Because of this, estimation of Δy is usually known as the *propagation* of uncertainty with which we know x_i through the data processing algorithm.

Uncertainty propagation: an equivalent formulation. By definition of the measurement error, we have

$$x_i = \widetilde{x}_i - \Delta x_i.$$

Thus, for the desired estimation error Δy, we get the following formula:

$$\Delta y = \widetilde{y} - y = f(\widetilde{x}_1, \ldots, \widetilde{x}_n) - f(\widetilde{x}_1 - \Delta x_1, \ldots, \widetilde{x}_n - \Delta x_n).$$

Our goal is to transform the available information about Δx_i into the information about the desired estimation error Δy.

What do we know about Δx_i: ideal case. Ideally, for each i, we should know:

• which values of Δx_i are possible, and
• how frequently can we expect each of these possible values.

In other words, in the ideal case, for every i, we should know the probability distribution of the corresponding measurement error.

Ideal case: how to estimate Δy? In some situations, we have analytical expressions for estimating Δy.

In other situations, since we know the exact probability distributions corresponding to all i, we can use *Monte-Carlo simulations* to estimate Δy. Namely, several times $\ell = 1, 2, \ldots, L$, we:

• simulate the values $\Delta x_i^{(\ell)}$ according to the known distribution of Δx_i, and
• estimate

$$\Delta y^{(\ell)} = \widetilde{y} - f(\widetilde{x}_1 - \Delta x_1^{(\ell)}, \ldots, \widetilde{x}_n - \Delta x_n^{(\ell)}).$$

Since the values $\Delta x_i^{(\ell)}$ have the exact same distribution as Δx_i, the computed values $\Delta y^{(\ell)}$ are a sample from the same distribution as Δy. Thus, from this sample

$$\Delta y^{(1)}, \ldots, \Delta y^{(L)},$$

we can find all necessary characteristics of the corresponding Δy-probability distribution.

What if we only have partial information about the probability distributions? In practice, we rarely full full information about the probabilities of different values of the measurement errors Δx_i, we only have partial information about these probabilities; see, e.g., [82]. In such situations, it is necessary to transform this partial information into the information about Δy.

What partial information do we have? What type of information can we know about Δx_i? To answer this question, let us take into account that the ultimate goal of all these estimations is to make a decision:

• when we estimate tomorrow's temperature, we make a decision of what to wear, or, in agriculture, a decision on whether to start planting the field;

- when we estimate the amount of oil, we make a decision whether to start drilling
  right now or to wait until the oil prices will go up since at present, the expected
  amount of oil is too large enough to justify the drilling expenses.

According to decision theory results (see, e.g., [19, 46, 59, 84]), a rational decision maker always selects an alternative that maximizes the expected value of some
objective function $u(x)$ – known as *utility*. From this viewpoint, it is desirable to select
characteristics of the probability distribution that help us estimate this expected value
– and thus, help us estimate the corresponding utility.

For each quantity x_i, depending on the measurement error Δx_i, we have different
values of the utility $u(\Delta x_i)$. For example:

- If we overestimate the temperature and start planting the field too early, we may
  lose some crops and thus, lose potential profit.
- If we start drilling when the actual amount of oil is too low – or, vie versa, do not
  start drilling when there is actually enough of oil – we also potentially lose money.

The measurement errors Δx_i are usually reasonably small. So, we can expand the
expression for the utility $u(\Delta x_i)$ in Taylor series and keep only the first few terms in
this expansion:

$$u(\Delta x_i) \approx u(0) + u_1 \cdot \Delta x_i + u_2 \cdot (\Delta x_i)^2 + \cdots + u_k \cdot (\Delta x_i)^k,$$

where the coefficients u_i are uniquely determined by the corresponding utility function $u(\Delta x_i)$. By taking the expected value $E[\cdot]$ of both sides of the above equality,
we conclude that

$$E[u(\Delta x_i)] \approx u(0) + u_1 \cdot E[\Delta x_i] + u_2 \cdot E[(\Delta x_i)^2] + \cdots + u_k \cdot E[(\Delta x_i)^k].$$

Thus, to compute the expected utility, it is sufficient to know the first few moments

$$E[\Delta x_i], \quad E[(\Delta x_i)^2], \ldots, E[(\Delta x_i)^k]$$

of the corresponding distribution.

From this viewpoint, a reasonable way to describe a probability distribution is via
its first few moments. This is what we will consider in this section.

**From the computational viewpoint, it is convenient to use cumulants, not
moments themselves**. From the computational viewpoint, in computational statistics, it is often more convenient to use not the moments themselves but their combinations called *cumulants*; see, e.g., [97]. A general mathematical definition of the
kth order cumulant κ_{in} of a random variable Δx_i is that it is a coefficient in the Taylor
expansion of the logarithm of the *characteristic function*

$$\chi_i(\omega) \overset{\text{def}}{=} E[\exp(i \cdot \omega \cdot \Delta x_i)]$$

(where $i \overset{\text{def}}{=} \sqrt{-1}$) in terms of ω:

$$\ln(E[\exp(i \cdot \omega \cdot \Delta x_i)]) = \sum_{k=1}^{\infty} \kappa_{ik} \cdot \frac{(i \cdot \omega)^k}{k!}.$$

It is known that the kth order cumulant can be described in terms of the moments up to order k; for example:

- κ_{i1} is simply the expected value, i.e., the first moment;
- κ_{i2} is minus variance;
- κ_{i3} and κ_{i4} are related to skewness and excess, etc.

The convenient thing about cumulants (as opposed to moments) is that when we add two independent random variables, their cumulants also add:

- the expected value of the sum of two independence random variables is equal to the sum of their expected values (actually, for this case, we do not even need independence, in other cases we do);
- the variance of the sum of two independent random variables is equal to the sum of their variance, etc.

In addition to this important property, kth order cumulants have many of the same properties of the kth order moments. For example:

- if we multiply a random variable by a constant c,
- then both its kth order moment and its kth order cumulant will multiply by c^k.

Usually, we know the cumulants only approximately. Based on the above explanations, a convenient way to describe each measurement uncertainty Δx_i is by describing the corresponding cumulants κ_{ik}.

The value of these cumulants also come from measurements. As a result, we usually know them only approximately, i.e., have an approximate value $\widetilde{\kappa}_{ik}$ and the upper bound Δ_{ik} on the corresponding inaccuracy:

$$|\kappa_{ik} - \widetilde{\kappa}_{ik}| \le \Delta_{ik}.$$

In this case, the only information that we have about the actual (unknown) values κ_{ik} is that each of these values belongs to the corresponding interval

$$[\underline{\kappa}_{ik}, \overline{\kappa}_{ik}],$$

where

$$\underline{\kappa}_{ik} \stackrel{\text{def}}{=} \widetilde{\kappa}_{ik} - \Delta_{ik}$$

and

$$\overline{\kappa}_{ik} \stackrel{\text{def}}{=} \widetilde{\kappa}_{ik} + \Delta_{ik}.$$

Thus, we arrive at the following formulation of the uncertainty propagation problem.

Uncertainty propagation: formulation of the problem. We know:

- an algorithm

$$f(x_1, \ldots, x_n),$$

- the measurement results

$$\widetilde{x}_1, \ldots, \widetilde{x}_n,$$

  and
- for each i from 1 to n, we know intervals

$$[\underline{\kappa}_{ik}, \overline{\kappa}_{ik}] = [\widetilde{\kappa}_{ik} - \Delta_{ik}, \widetilde{\kappa}_{ik} + \Delta_{ik}]$$

  that contain the actual (unknown) cumulants κ_{ik} of the measurement errors

$$\Delta x_i = \widetilde{x}_i - x_i.$$

Based on this information, we need to compute the range

$$[\underline{\kappa}_k, \overline{\kappa}_k]$$

of possible values of the cumulants κ_k corresponding to

$$\Delta y = f(\widetilde{x}_1, \ldots, \widetilde{x}_n) - f(x_1, \ldots, x_n) = f(\widetilde{x}_1, \ldots, \widetilde{x}_n) - f(\widetilde{x}_1 - \Delta x_1, \ldots, \widetilde{x}_n - \Delta x_n).$$

3.4.3   Existing Algorithms for Uncertainty Propagation and Their Limitations

Usually, measurement errors are relatively small. As we have mentioned, in most practical cases, the measurement error is relatively small. So, we can safely ignore terms which are quadratic (or of higher order) in terms of the measurement errors. For example:

- if we measure something with 10% accuracy,
- then the quadratic terms are of order 1%, which is definitely much less than 1%.

Thus, to estimate Δy, we can expand the expression for Δy in Taylor series and keep only linear terms in this expansion. Here, by definition of the measurement error, we have

$$x_i = \widetilde{x}_i - \Delta x_i,$$

thus

$$\Delta y = f(\widetilde{x}_1, \ldots, \widetilde{x}_n) - f(\widetilde{x}_1 - \Delta x_1, \ldots, \widetilde{x}_n - \Delta x_n).$$

Expanding the right-hand side in Taylor series and keeping only linear terms in this expansion, we conclude that

$$\Delta y = \sum_{i=1}^{n} c_i \cdot \Delta x_i,$$

where c_i is the value of the ith partial derivative $\dfrac{\partial f}{\partial x_i}$ at a point $(\tilde{x}_1, \ldots, \tilde{x}_n)$:

$$c_i \overset{\text{def}}{=} \frac{\partial f}{\partial x_i}(\tilde{x}_1, \ldots, \tilde{x}_n).$$

Let us derive explicit formulas for $\underline{\kappa}_k$ and $\overline{\kappa}_k$. Let us assume that we know the coefficients c_i.

Due to the above-mentioned properties of cumulants, if κ_{ik} is the kth cumulant of Δx_i, then the kth cumulant of the product $c_i \cdot \Delta x_i$ is equal to

$$(c_i)^k \cdot \kappa_{ik}.$$

In its turn, the kth order cumulant κ_k for the sum Δy of these products is equal to the sum of the corresponding cumulants:

$$\kappa_k = \sum_{i=1}^{n} (c_i)^k \cdot \kappa_{ik}.$$

We can represent each (unknown) cumulant κ_{ik} as the difference

$$\kappa_{ik} = \tilde{\kappa}_{ik} - \Delta \kappa_{ik},$$

where

$$\Delta \kappa_{ik} \overset{\text{def}}{=} \tilde{\kappa}_{ik} - \kappa_{ik}$$

is bounded by the known value Δ_{ik}:

$$|\Delta \kappa_{ik}| \leq \Delta_{ik}.$$

Substituting the above expression for κ_{ik} into the formula for κ_k, we conclude that

$$\kappa_k = \tilde{\kappa}_k - \Delta \kappa_k,$$

where we denoted

$$\tilde{\kappa}_k \overset{\text{def}}{=} \sum_{i=1}^{n} (c_i)^k \cdot \tilde{\kappa}_k$$

and

$$\Delta \kappa_k \stackrel{\text{def}}{=} \sum_{i=1}^{n} (c_i)^k \cdot \Delta \kappa_{ik}.$$

The value $\widetilde{\kappa}_k$ is well defined. The value $\Delta \kappa_k$ depends on the approximation errors $\Delta \kappa_{ik}$. To find the set of possible values κ_k, we thus need to find the range of possible values of $\Delta \kappa_k$.

This value is the sum of n independent terms, independent in the sense that each of them depends only on its own variable $\Delta \kappa_{ik}$. So, the sum attains its largest values when each of the terms

$$(c_i)^k \cdot \Delta \kappa_{ik}$$

is the largest.

- When $(c_i)^k > 0$, the expression $(c_i)^k \cdot \Delta \kappa_{ik}$ is an increasing function of $\Delta \kappa_{ik}$, so it attains its largest possible value when $\Delta \kappa_{ik}$ attains its largest possible value Δ_{ik}. The resulting largest value of this term is

$$(c_i)^k \cdot \Delta_{ik}.$$

- When $(c_i)^k < 0$, the expression $(c_i)^k \cdot \Delta \kappa_{ik}$ is a decreasing function of $\Delta \kappa_{ik}$, so it attains its largest possible value when $\Delta \kappa_{ik}$ attains its smallest possible value $-\Delta_{ik}$. The resulting largest value of this term is

$$-(c_i)^k \cdot \Delta_{ik}.$$

Both cases can be combined into a single expression $|(c_i)^k| \cdot \Delta_{ik}$ if we take into account that:

- when $(c_i)^k > 0$, then $|(c_i)^k| = (c_i)^k$, and
- when $(c_i)^k < 0$, then $|(c_i)^k| = -(c_i)^k$.

Thus, the largest possible value of $\Delta \kappa_k$ is equal to

$$\Delta_k \stackrel{\text{def}}{=} \sum_{i=1}^{n} |(c_i)^k| \cdot \Delta_{ik}.$$

Similarly, we can show that the smallest possible value of $\Delta \kappa_k$ is equal to $-\Delta_k$. Thus, we arrive at the following formulas for computing the desired range $[\underline{\kappa}_k, \overline{\kappa}_k]$.

Explicit formulas for $\underline{\kappa}_k$ and $\overline{\kappa}_k$. Here, $\underline{\kappa}_k = \widetilde{\kappa}_k - \Delta_k$ and $\overline{\kappa}_k = \widetilde{\kappa}_k + \Delta_k$, where

$$\widetilde{\kappa}_k = \sum_{i=1}^{n} (c_i)^k \cdot \widetilde{\kappa}_k$$

and

$$\Delta_k = \sum_{i=1}^{n} |(c_i)^k| \cdot \Delta_{ik}.$$

A resulting straightforward algorithm. The above formulas can be explicitly used to estimate the corresponding quantities. The only remaining question is how to estimate the corresponding values c_i of the partial derivatives.

- When $f(x_1, \ldots, x_n)$ is an explicit expression, we can simply differentiate the function f and get the values of the corresponding derivatives.
- In more complex cases, e.g., when the algorithm $f(x_1, \ldots, x_n)$ is given as a proprietary black box, we can compute all the values c_i by using numerical differentiation:

$$c_i \approx \frac{f(\widetilde{x}_1, \ldots, \widetilde{x}_{i-1}, \widetilde{x}_i + \varepsilon_i, \widetilde{x}_{i+1}, \ldots, \widetilde{x}_n) - \widetilde{y}}{\varepsilon_i}$$

for some small ε_i.

Main limitation of the straightforward algorithm: it takes too long. When $f(x_1, \ldots, x_n)$ is a simple expression, the above straightforward algorithm is very efficient.

However, in many cases – e.g., with weather prediction or oil exploration – the corresponding algorithm $f(x_1, \ldots, x_n)$ is very complex and time-consuming,

- requiring hours of computation on a high performance computer,
- while processing thousands of data values x_i.

In such situations, the above algorithm requires $n + 1$ calls to the program that implements the algorithm $f(x_1, \ldots, x_n)$:

- one time to compute

$$\widetilde{y} = f(\widetilde{x}_1, \ldots, \widetilde{x}_n),$$

and then
- n times to compute n values

$$f(\widetilde{x}_1, \ldots, \widetilde{x}_{i-1}, \widetilde{x}_i + \varepsilon_i, \widetilde{x}_{i+1}, \ldots, \widetilde{x}_n)$$

needed to compute the corresponding partial derivatives c_i.

When each call to f takes hours, and we need to make thousands of such class, the resulting computation time is in years.

This makes the whole exercise mostly useless: when it takes hours to predict the weather, no one will wait more than a year to check how accurate is this prediction. It is therefore necessary to have faster methods for uncertainty propagation.

Much faster methods exist for moments (and cumulants) of even order k. For all k, the computation of the value

$$\kappa_k = \sum_{i=1}^{n} (c_i)^k \cdot \widetilde{\kappa}_{ik}$$

can be done much faster, by using the following Monte-Carlo simulations.

Several times $\ell = 1, 2, \ldots, L$, we:

- simulate the values $\Delta x_i^{(\ell)}$ according to some distribution of Δx_i with the given value $\widetilde{\kappa}_{ik}$, and
- estimate

$$\Delta y^{(\ell)} = \widetilde{y} - f(\widetilde{x}_1 - \Delta x_1^{(\ell)}, \ldots, \widetilde{x}_n - \Delta x_n^{(\ell)}).$$

One can show that in this case, the kth cumulant of the resulting distribution for $\Delta y^{(\ell)}$ is equal to exactly the desired value

$$\kappa_k = \sum_{i=1}^{n} (c_i)^k \cdot \widetilde{\kappa}_{ik}.$$

Thus, by computing the sample moments of the sample

$$\Delta y^{(1)}, \ldots, \Delta y^{(L)},$$

we can find the desired kth order cumulant.

For example, for $k = 2$, when the cumulant is the variance, we can simply use normal distributions with a given variance.

The main advantage of the Monte-Carlo method is that its accuracy depends only on the number of iterations: its uncertainty decreases with L as $1/\sqrt{L}$; see, e.g., [97]. Thus, for example:

- to get the moment with accuracy 20% ($= 1/5$),
- it is sufficient to run approximately 25 simulations, i.e., approximately 25 calls to the algorithm f; this is much much faster than thousands of iterations needed to perform the straightforward algorithm.

For even k, the value $(c_i)^k$ is always non-negative, so

$$|(c_i)^k| = (c_i)^k,$$

and the formula for Δ_k get a simplified form

$$\Delta_k = \sum_{i=1}^{n} (c_i)^k \cdot \Delta_{ik}.$$

This is exactly the same form as for $\widetilde{\kappa}_k$, so we can use the same Monte-Carlo algorithm to estimate Δ_k – the only difference is that now, we need to use distributions of Δx_i with the kth cumulant equal to Δ_{ik}.

Specifically, several times $\ell = 1, 2, \ldots, L$, we:

- simulate the values $\Delta x_i^{(\ell)}$ according to some distribution of Δx_i with the value Δ_{ik} of the kth cumulant, and
- estimate

$$\Delta y^{(\ell)} = \widetilde{y} - f(\widetilde{x}_1 - \Delta x_1^{(\ell)}, \ldots, \widetilde{x}_n - \Delta x_n^{(\ell)}).$$

One can show that in this case, the kth cumulant of the resulting distribution for $\Delta y^{(\ell)}$ is equal to exactly the desired value

$$\Delta_k = \sum_{i=1}^{n} (c_i)^k \cdot \Delta_{ik}.$$

Thus, by computing the sample moments of the sample

$$\Delta y^{(1)}, \ldots, \Delta y^{(L)},$$

we can find the desired bound Δ_k on the kth order cumulant.

Odd order moments (such as skewness) remain a computational problem. For odd k, we can still use the same Monte-Carlo method to compute the value $\widetilde{\kappa}_k$.

However, we can no longer use this method to compute the bound Δ_k on the kth cumulant, since for odd k, we no longer have the equality

$$|(c_i)^k| = (c_i)^k.$$

What we plan to do. We will show that the use of (hypothetical) negative probabilities enables us to attain the same speed up for the case of odd k as we discussed above for the case of even orders.

3.4.4 Analysis of the Problem and the Resulting Negative-Probability-Based Fast Algorithm for Uncertainty Quantification

Why the Monte-Carlo method works for variances? The possibility to use normal distributions to analyze the propagation of variances

$$V = \sigma^2$$

comes from the fact that if we have n independent random variables Δx_i with variances

$$V_i = \sigma_i^2,$$

then their linear combination

$$\Delta y = \sum_{i=1}^{n} c_i \cdot \Delta x_i$$

is also normally distributed, with variance

$$V = \sum_{i=1}^{n} (c_i)^2 \cdot V_i,$$

– and this is exactly how we want to relate the variance (2nd order cumulant) of Δy with the variances V_i of the inputs.

Suppose that we did not know that the normal distribution has this property. How would we then be able to find a distribution $\rho_1(x)$ that satisfies this property? Let us consider the simplest case of this property, when

$$V_1 = \cdots = V_n = 1.$$

In this case, the desired property has the following form:

- if n independent random variables $\Delta x_1, \ldots, \Delta x_n$ have exactly the same distribution, with variance 1,
- then their linear combination

$$\Delta y = \sum_{i=1}^{n} c_i \cdot \Delta x_i$$

has the same distribution, but re-scaled, with variance

$$V = \sum_{i=1}^{n} (c_i)^2.$$

Let $\rho_1(x)$ denote the desired probability distribution, and let

$$\chi_1(\omega) = E[\exp(i \cdot \omega \cdot \Delta x_1)]$$

be the corresponding characteristic function. Then, for the product $c_i \cdot \Delta x_i$, the characteristic function has the form

$$E[\exp(i \cdot \omega \cdot (c_i \cdot \Delta x_1))].$$

By re-arranging multiplications, we can represent this same expression as

$$E[\exp(i \cdot (\omega \cdot c_i) \cdot \Delta x_1)],$$

i.e., as $\chi_1(c_i \cdot \omega)$.

For the sum of several independent random variables, the characteristic function is equal to the product of characteristic functions (see, e.g., [97]); thus, the characteristic function of the sum

$$\sum_{i=1}^{n} c_i \cdot \Delta x_i$$

has the form

$$\chi_1(c_1 \cdot \omega) \cdot \cdots \cdot \chi_1(c_n \cdot \omega).$$

We require that this sum be distributed the same way as Δx_i, but with a larger variance. When we multiply a variable by c, its variable increases by a factor of c^2. Thus, to get the distribution with variance

$$V = \sum_{i=1}^{n} (c_i)^2,$$

we need to multiply the variable Δx_i by a factor of

$$c = \sqrt{\sum_{i=1}^{n} (c_i)^2}.$$

For a variable multiplied by this factor, the characteristic function has the form

$$\chi_1(c \cdot \omega).$$

By equating the two characteristic functions, we get the following functional equation:

$$\chi_1(c_1 \cdot \omega) \cdot \cdots \cdot \chi_1(c_n \cdot \omega) = \chi_1 \left(\sqrt{\sum_{i=1}^{n} (c_i)^2} \cdot \omega \right).$$

In particular, for $n = 2$, we conclude that

$$\chi_1(c_1 \cdot \omega) \cdot \chi_1(c_2 \cdot \omega) = \chi_1 \left(\sqrt{(c_1)^2 + (c_2)^2} \cdot \omega \right).$$

This expression can be somewhat simplified if we take logarithms of both sides. Then products turn to sums, and for the new function

$$\ell(\omega) \overset{\text{def}}{=} \ln(\chi_1(\omega)),$$

we get the equation

$$\ell(c_1 \cdot \omega) + \ell(c_2 \cdot \omega) = \ell\left(\sqrt{(c_1)^2 + (c_2)^2} \cdot \omega\right).$$

This equation can be further simplified if we consider an auxiliary function

$$F(\omega) \overset{\text{def}}{=} \ell(\sqrt{\omega}),$$

for which

$$\ell(x) = F(x^2).$$

Substituting the expression for $\ell(x)$ in terms of $F(x)$ into the above formula, we conclude that

$$F((c_1)^2 \cdot \omega^2) + F((c_2)^2 \cdot \omega^2) = F(((c_1)^2 + (c_2)^2) \cdot \omega^2).$$

One can easily check that for every two non-negative numbers a and b, we can take

$$\omega = 1, c_1 = \sqrt{a}, \text{ and } c_2 = \sqrt{b},$$

and thus turn the above formula into

$$F(a) + F(b) = F(a + b).$$

It is well known (see, e.g., [1]) that every measurable solution to this functional equation has the form

$$F(a) = K \cdot a$$

for some constant K. Thus,

$$\ell(\omega) = F(\omega^2) = K \cdot \omega^2.$$

Here,

$$\ell(\omega) = \ln(\chi_1(\omega)),$$

hence

$$\chi_1(\omega) = \exp(\ell(\omega)) = \exp(K \cdot \omega^2).$$

Based on the characteristic function, we can reconstruct the original probability density function $\rho_1(x)$. Indeed, from the purely mathematical viewpoint, the characteristic function

$$\chi(\omega) = E[\exp(i \cdot \omega \cdot \Delta x_1)] = \int \exp(i \cdot \omega \cdot \Delta x_1) \cdot \rho_1(\Delta x_1) \, d(\Delta x_1)$$

is nothing else but the Fourier transform of the probability density function $\rho_1(\Delta x_1)$. We can therefore always reconstruct the original probability density function by applying the inverse Fourier transform to the characteristic function.

For

$$\chi_1(\omega) = \exp(K \cdot \omega^2),$$

the inverse Fourier transform leads to the usual formula of the normal distribution, with

$$K = -\sigma^2.$$

Can we apply the same idea to odd k? Our idea us to use Monte-Carlo methods for odd k, to speed up the computation of the value

$$\Delta_k = \sum_{i=1}^{n} |(c_i)^k| \cdot \Delta_{ik}.$$

What probability distribution $\rho_1(x)$ can we use to do it?

Similar to the above, let us consider the simplest case when

$$\Delta_{1k} = \cdots = \Delta_{nk} = 1.$$

In this case, the desired property of the probability distribution takes the following form:

• if n independent random variables

$$\Delta x_1, \ldots, \Delta x_n$$

have exactly the same distribution $\rho_1(x)$, with kth cumulant equal to 1,
• then their linear combination

$$\Delta y = \sum_{i=1}^{n} c_i \cdot \Delta x_i$$

has the same distribution, but re-scaled, with the kth order cumulant equal to

$$\sum_{i=1}^{n} |c_i|^k.$$

Let $\rho_1(x)$ denote the desired probability distribution, and let

$$\chi_1(\omega) = E[\exp(i \cdot \omega \cdot \Delta x_1)]$$

be the corresponding characteristic function. Then, as we have shown earlier, for the product $c_i \cdot \Delta x_i$, the characteristic function has the form $\chi_1(c_i \cdot \omega)$. For the sum

$$\sum_{i=1}^{n} c_i \cdot \Delta x_i,$$

the characteristic function has the form

$$\chi_1(c_1 \cdot \omega) \cdot \cdots \cdot \chi_1(c_n \cdot \omega).$$

We require that this sum be distributed the same way as Δx_i, but with a larger kth order cumulant. As we have mentioned:

- when we multiply a variable by c,
- its kth order cumulant increases by a factor of c^k.

Thus, to get the distribution with the value

$$\sum_{i=1}^{n} |c_i|^k,$$

we need to multiply the variable Δx_i by a factor of

$$c = \sqrt[k]{\sum_{i=1}^{n} |c_i|^k}.$$

For a variable multiplied by this factor, the characteristic function has the form

$$\chi_1(c \cdot \omega).$$

By equating the two characteristic functions, we get the following functional equation:

$$\chi_1(c_1 \cdot \omega) \cdot \cdots \cdot \chi_1(c_n \cdot \omega) = \chi_1 \left(\sqrt[k]{\sum_{i=1}^{n} |c_i|^k} \cdot \omega \right).$$

In particular, for $n = 2$, we conclude that

$$\chi_1(c_1 \cdot \omega) \cdot \chi_1(c_2 \cdot \omega) = \chi_1 \left(\sqrt[k]{|c_1|^k + |c_2|^k} \cdot \omega \right).$$

This expression can be somewhat simplified if we take logarithms of both sides. Then products turn to sums, and for the new function

$$\ell(\omega) \stackrel{\text{def}}{=} \ln(\chi_1(\omega)),$$

we get the equation

$$\ell(c_1 \cdot \omega) + \ell(c_2 \cdot \omega) = \ell\left(\sqrt[k]{(|c_1|^k + |c_2|^k} \cdot \omega\right).$$

This equation can be further simplified if we consider an auxiliary function

$$F(\omega) \stackrel{\text{def}}{=} \ell(\sqrt[k]{\omega}),$$

for which

$$\ell(x) = F(x^k).$$

Substituting the expression for $\ell(x)$ in terms of $F(x)$ into the above formula, we conclude that

$$F(|c_1|^k \cdot \omega^k) + F(|c_2|^k \cdot \omega^k) = F((|c_1|^k + |c_2|^k) \cdot \omega^k).$$

One can easily check that for every two non-negative numbers a and b, we can take

$$\omega = 1, c_1 = \sqrt[k]{a}, \text{ and } c_2 = \sqrt[k]{b}$$

and thus get

$$F(a) + F(b) = F(a + b).$$

As we have already shown, this leads to

$$F(a) = K \cdot a$$

for some constant K. Thus,

$$\ell(\omega) = F(\omega^k) = K \cdot \omega^k.$$

Here,

$$\ell(\omega) = \ln(\chi_1(\omega)),$$

hence

$$\chi_1(\omega) = \exp(\ell(\omega)) = \exp(K \cdot \omega^k).$$

Case of $k = 1$ leads to a known efficient method. For $k = 1$, the above characteristic function has the form

$$\exp(-K \cdot |\omega|).$$

By applying the inverse Fourier transform to this expression, we get the Cauchy distribution, with probability density

$$\rho_1(x) = \frac{1}{\pi \cdot K} \cdot \frac{1}{1 + \dfrac{x^2}{K^2}}.$$

Monte-Carlo methods based on the Cauchy distribution indeed lead to efficient estimation of first order uncertainty – e.g., bounds on mean; see, e.g., [34].

What about larger odd values k? Alas, for $k \geq 3$, we have a problem:

- when we apply the inverse Fourier transform to the characteristic function

$$\exp(-|K| \cdot |\omega|^k),$$

- the resulting function $\rho_1(\Delta x_1)$ takes negative values for some x, and thus, cannot serve as a usual probability density function; see, e.g., [90].

However:

- if negative probabilities are physically possible,
- then we can indeed use the same idea to speed up computation of Δ_k for odd values

$$k \geq 3.$$

If negative probabilities are physically possible, then we can speed up uncertainty propagation – namely, computation of Δ_k. If negative probabilities are indeed physically possible, then we can use the following algorithm to speed up the computation of Δ_k.

Let us assume that we are able to simulate a "random" variable η whose (sometimes negative) probability density function $\rho_1(x)$ is the inverse Fourier transform of the function

$$\chi_1(\omega) = \exp(-|\omega|^k).$$

We will use the corresponding "random" number generator for each variable x_i and for each iteration $\ell = 1, 2, \ldots, L$. The corresponding value will be denoted by $\eta_i^{(\ell)}$.

The value $\eta_i^{(\ell)}$ will corresponds to the value of the kth cumulant equal to 1. To simulate a random variable corresponding to parameter Δ_{ik}, we use

$$(\Delta_{ik})^{1/k} \cdot \eta_i^{(\ell)}.$$

Thus, we arrive at the following algorithm:

Several times $\ell = 1, 2, \ldots, L$, we:

- simulate the values $\Delta x_i^{(\ell)}$ as

$$(\Delta_{ik})^{1/k} \cdot \eta_i^{(\ell)},$$

and

- estimate

$$\Delta y^{(\ell)} = \tilde{y} - f(\tilde{x}_1 - \Delta x_1^{(\ell)}, \ldots, \tilde{x}_n - \Delta x_n^{(\ell)}).$$

One can show that in this case, the resulting distribution for $\Delta y^{(\ell)}$ has the same distribution as η multiplied by the kth root of the desired value

$$\Delta_k = \sum_{i=1}^{n} (c_i)^k \cdot \Delta_{ik}.$$

Thus, by computing the corresponding characteristic of the sample

$$\Delta y^{(1)}, \ldots, \Delta y^{(L)},$$

we can find the desired bound Δ_k on the kth order cumulant.

So, we can indeed use fast Monte-Carlo methods to estimate both values $\widetilde{\kappa}_k$ and Δ_k – and thus, to speed up uncertainty propagation.

Chapter 4
Towards a Better Understandability of Uncertainty-Estimating Algorithms

In this chapter, we explain how to make uncertainty-estimating algorithms easier to understand. We start with the case of interval uncertainty. For this case, in Sect. 4.1, we provide an intuitive explanation for different types of solutions, and in Sect. 4.2, we provide an intuitive explanation for seemingly counter-intuitive methods for solving the corresponding problem. In Sect. 4.3, we consider the case of probabilistic uncertainty. For this case, we explain why it is reasonable to consider mixtures of probability distributions. In Sect. 4.4, we consider the case of fuzzy uncertainty; we explain why seemingly non-traditional fuzzy ideas are, in effect, very similar to the ideas from traditional logic.

After analyzing specific types of uncertainty, in Sects. 4.5 and 4.6, we consider the general case. Specifically, in Sect. 4.5, we explain the ubiquity of linear dependencies, and in Sect. 4.6, we explain the consequences of deviation from linearity.

4.1 Case of Interval Uncertainty: Practical Need for Algebraic (Equality-Type) Solutions of Interval Equations and for Extended-Zero Solutions

One of the main problems in interval computations is solving systems of equations under interval uncertainty. Usually, interval computation packages consider united, tolerance, and control solutions. In this section, we explain the practical need for *algebraic* (equality-type) solutions, when we look for solutions for which both sides are equal. In situations when such a solution is not possible, we provide a justification for extended-zero solutions, in which we ignore intervals of the type $[-a, a]$.

Results presented in this section first appeared in [15].

© Springer International Publishing AG, part of Springer Nature 2018   97
A. Pownuk and V. Kreinovich, *Combining Interval, Probabilistic, and Other Types of Uncertainty in Engineering Applications*, Studies in Computational Intelligence 773, https://doi.org/10.1007/978-3-319-91026-0_4

4.1.1 Practical Need for Solving Interval Systems of Equations: What Is Known

Need for data processing. In many practical situations, we are interested in the values of quantities $y_1, \ldots, y_m$ which are difficult – or even impossible – to measure directly. For example, we can be interested in a distance to a faraway star or in tomorrow's temperature at a certain location.

Since we cannot measure these quantities directly, to estimate these quantities we must:

- find easier-to-measure quantities $x_1, \ldots, x_n$ which are related to y_i by known formulas $y_i = f_i(x_1, \ldots, x_n)$,
- measure these quantities x_j, and
- use the results $\tilde{x}_j$ of measuring the quantities x_j to compute the estimates for y_i:

$$\tilde{y}_i = f(\tilde{x}_1, \ldots, \tilde{x}_n).$$

Computation of these estimates is called *indirect measurement* or *data processing*.

Need for data processing under uncertainty. Measurements are never 100% accurate. Hence, the measurement result $\tilde{x}_j$ is, in general, different from the actual (unknown) value x_j of the corresponding quantity; in other words, the measurement errors $\Delta x_j \stackrel{\text{def}}{=} \tilde{x}_j - x_j$ are, in general, different from 0.

Because of the non-zero measurement errors, the estimates $\tilde{y}_i$ are, in general, different from the desired values y_i. It is therefore desirable to know how accurate are the resulting estimates.

Need for interval uncertainty and interval computations. The manufacturer of the measuring instrument usually provides us with an upper bound Δ_j on the measurement error: $|\Delta x_j| \leq \Delta_j$; see, e.g., [82]. If no such upper bound is known, i.e., if the reading of the instrument can be as far away from the actual value as possible, then this is not a measuring instrument, this is a wild-guess-generator.

Sometimes, we also know the probabilities of different values Δx_j within this interval; see, e.g., [82, 97]. However, in many practical situations, the upper bound is the only information that we have [82]. In this case, after we know the result $\tilde{x}_j$ of measuring x_j, the only information that we have about the actual (unknown) value x_j is that this value belongs to the interval $[\underline{x}_j, \overline{x}_j]$, where $\underline{x}_j \stackrel{\text{def}}{=} \tilde{x}_j - \Delta_j$ and $\overline{x}_j \stackrel{\text{def}}{=} \tilde{x}_j + \Delta_j$.

In this case, the only thing that we can say about each value $y_i = f_i(x_1, \ldots, x_n)$ is that this value belongs to the range

$$\{f_i(x_1, \ldots, x_n) : x_1 \in [\underline{x}_1, \overline{x}_1], \ldots, x_n \in [\underline{x}_n, \overline{x}_n]\}.$$

Computation of this range is one of the main problems of *interval computations*; see, e.g., [26, 58].

Sometimes, we do not know the exact dependence. The above text described an ideal case, when we know the exact dependence $y_i = f_i(x_1, \ldots, x_n)$ between the desired quantities y_i and the easier-to-measure quantities x_j. In practice, often, we do not know the exact dependence. Instead, we know that the dependence belongs to a finite-parametric *family* of dependencies, i.e., that

$$y_i = f_i(x_1, \ldots, x_n, a_1, \ldots, a_k)$$

for some parameters $a_1, \ldots, a_k$.

For example, we may know that y_i is a linear function of the quantities x_j, i.e., that $y_i = c_i + \sum_{j=1}^{n} c_{ij} \cdot x_j$ for some coefficients c_i and c_{ij}.

The presence of these parameters complicates the corresponding data processing problem. Depending on what we know about the parameters, we have different situations.

Simplest situation, when we know the exact values of all the parameters. The simplest situation is when we know the exact values of these parameters. In this case, the dependence of y_i on x_j is known, and we have the same problem of computing the range as before.

Specific case: control solution. Sometimes, not only we *know* the values a_ℓ of these parameters, but we can also *control* these values, by setting them to any values within certain intervals $[\underline{a}_\ell, \overline{a}_\ell]$. By setting the appropriate values of the parameters, we can change the values y_i. This possibility naturally leads to the following problem:

- we would like the values y_i to be within some given ranges $[\underline{y}_i, \overline{y}_i]$; for example, we would like the temperature to be within a comfort zone;
- in this case, it is desirable to find the range of possible values of x_j for which, by applying appropriate controls $a_i \in [\underline{a}_\ell, \overline{a}_\ell]$, we can place the values y_i within these intervals.

In the degenerate case, when all the intervals for y_i and a_ℓ are just points, this means solving the system of equations $y = f(x, a)$, where we denoted $y \stackrel{\text{def}}{=} (y_1, \ldots, y_m)$, $x \stackrel{\text{def}}{=} (x_1, \ldots, x_n)$, and $a \stackrel{\text{def}}{=} (a_1, \ldots, a_k)$. From this viewpoint, the above problem can be viewed as an interval generalization of the problem of solving a system of equations, or, informally, as a problem of solving the corresponding interval system of equations.

The set X of all appropriate values $x = (x_1, \ldots, x_n)$ can be formally described as

$$X = \{x : \text{ for some } a_\ell \in [\underline{a}_\ell, \overline{a}_\ell], f_i(x_1, \ldots, x_n, a_1, \ldots, a_k) \in [\underline{y}_i, \overline{y}_i] \text{ for all } i\}.$$

This set is known as the *control solution* to the corresponding interval system of equations [26, 96].

Situation when we need to find the parameters from the data. Sometimes, we know that the values a_i are the same for all the cases, but we do not know these values. These values must then be determined based on measurements: we measure x_j and y_i several times, and we find the values of the parameters a_ℓ that match all the measurement results.

Let us number all membership cycles by values $c = 1, \ldots, C$. After each cycle of measurements, we conclude that:

- the actual (unknown) value of $x_j^{(c)}$ is in the interval $[\underline{x}_j^{(c)}, \overline{x}_j^{(c)}]$ and
- the actual value of $y_i^{(c)}$ is in the interval $[\underline{y}_i^{(c)}, \overline{y}_i^{(c)}]$.

We want to find the set A of all the values a for which $y^{(c)} = f(x^{(c)}, a)$ for some $x^{(c)}$ and $y^{(c)}$:

$$A = \{a : \forall c \, \exists x_j^{(c)} \in [\underline{x}_j^{(c)}, \overline{x}_j^{(c)}] \, \exists y_i^{(c)} \in [\underline{y}_i^{(c)}, \overline{y}_i^{(c)}] \, (f(x^{(c)}, a) = y^{(c)})\}.$$

This set A is known as the *united solution* to the interval system of equations [26, 96].

Comment. To avoid confusion, it is worth mentioning that our notations are somewhat different from the notations used in [26, 96].

The main reason for this difference is that the main focus of this section is on the *motivations* for different types of solutions. As a result, we use the notations related to the meaning of the corresponding variables. In general, in our description, y denotes the desired (difficult-to-measure) quantities, x denote easier-to-measure quantities, and a denote parameters of the dependence between these quantities.

Within this general situation, we can have different problems.

- In some cases, we have some information about the parameters a, and we need to know the values x – this is the case of the control solution.
- In other practical situations, we have some information about the quantities x, and we need to know the values a – this is the case, e.g., for the united solution.

As a result, when we use our meaning-of-variables notations, sometimes x's are the unknowns, and sometimes a's are the unknowns.

Alternatively, if we were interested in actually solving the corresponding problems, it would be more appropriate to use different notations, in which, e.g., the unknown is always denoted by x and the known values are denoted by a – irrespective of the physical meaning of the corresponding variables. In these notations, the united solution would take a different form

$$X = \{x : \forall c \, \exists a_j^{(c)} \in [\underline{a}_j^{(c)}, \overline{a}_j^{(c)}] \, \exists y_i^{(c)} \in [\underline{y}_i^{(c)}, \overline{y}_i^{(c)}] \, (f(x, a^{(c)}) = y^{(c)})\}.$$

What can we do once we have found the range of possible values of a. Once we have found the set A of possible values of a, we can first find the range of possible values of y_i based on the measurement results, i.e., find the range

$$\{f_i(x_1, \ldots, x_n, a) : x_j \in [\underline{x}_j, \overline{x}_j] \text{ and } a \in A\}.$$

This is a particular case of the main problem of interval computations.

If we want to make sure that each value y_i lies within the given bounds $[\underline{y}_i, \overline{y}_i]$, then we must find the set X of possible values of x for which $f_i(x, a)$ is within these bounds for all possible values $a \in A$, i.e., the set

$$X = \{x : \forall a \in A \, \forall i \, (f_i(x, a) \in [\underline{y}_i, \overline{y}_i])\}.$$

This set is known as the *tolerance solution* to the interval system of equations [26, 96].

Sometimes, we know that the values a may change. In the previous text, we consider the situations when the values a_ℓ are either fixed forever, or can be changed by us. In practice, these values may change in an unpredictable way – e.g., if these parameters represent some physical processes that influence y_i's. We therefore do not know the exact values of these parameters, but what we do know is some a priori bounds on these values.

We may know bounds $[\underline{a}_\ell, \overline{a}_\ell]$ on each parameter, in which case the set A of all possible combinations $a = (a_1, \ldots, a_k)$ is simply a box:

$$A = [\underline{a}_1, \overline{a}_1] \times \ldots \times [\underline{a}_k, \overline{a}_k].$$

We may also have more general sets A – e.g., ellipsoids.

In this case, we can still solve the same two problems whose solutions we described above; namely:

- we can solve the main problem of interval computations – the problem of computing the range – and find the set Y of possible values of y;
- we can also solve the corresponding tolerance problem and find the set of values x that guarantee that each y_i is within the desired interval.

Is this all there is? There are also more complex problems (see, e.g., [96]), but, in a nutshell, most practical problems are either range estimation, or finding control, united, or tolerance solution. These are the problems solved by most interval computation packages [26, 58].

Is there anything else? In this section, we show that there is an important class of practical problems that does not fit into one of the above categories. To solve these practical problems, we need to use a different notion of a solution to interval systems of equations: the notion of an *algebraic* (equality-type) solution, the notion that has been previously proposed and theoretically analyzed [11, 12, 45, 62, 83, 94–96] but is not usually included in interval computations packages.

4.1.2 Remaining Problem of How to Find the Set A Naturally Leads to Algebraic (Equality-Type) Solutions to Interval System of Equations

Finding the set A: formulation of the problem. In the previous text, we assumed that when the values of the parameter a can change, we know the set A of possible values of the corresponding parameter vector. But how do we find this set?

What information we can use to find the set A. All the information about the real world comes from measurements – either directly from measurements, or by processing measurement results. The only relation between the parameters a and measurable quantities is the formula $y = f(x, a)$. Thus, to find the set A of possible values of a, we need to use measurements of x and y.

We can measure both x and y many times. As a result, we get:

- the set X of possible values of the vector x and
- the set Y of possible values of the vector y.

Both sets can be boxes, or they can be more general sets.

Based on these two sets X and Y, we need to find the set A.

In this problem, it is reasonable to assume that x and a are *independent* in some reasonable sense. Let us formulate this requirement in precise terms.

Independence: towards a formal definition. The notion of independence is well known in the probabilistic case, where it means that probability of getting a value $x \in X$ does not depend on the value $a \in A$: $P(x \mid a) = P(x \mid a')$ for all $a, a' \in A$. An interesting corollary of this definition is that, in spite of being formulated in a way that is asymmetric with respect to x and a, this definition is actually symmetric: one can prove that a is independent of x if and only if x is independent of a.

In the interval case, we do not know the probabilities, we only know which pairs (x, a) are possible and which are not. In other words, we have a *set $S \subseteq X \times A$* of possible pairs (x, a). It is natural to say that the values x and a are independent if the set of possible values of x does not depend on a. Thus, we arrive at the following definition.

Definition 4.1.1 Let $S \subseteq X \times A$ be a set.

- We say that a pair (x, a) is *possible* if $(x, a) \in S$.
- Let $x \in X$ and $a \in A$. We say that a value x is possible under a if $(x, a) \in S$. The set of possible-under-a values will be denoted by S_a.
- We say that the variables x and a are *independent* if $S_a = S_{a'}$ for all a, a' from the set A.

Proposition 4.1.1 *Variable x and a are independent if and only if S is a Cartesian product, i.e., $S = s_x \times s_a$ for some $s_x \subseteq X$ and $s_a \subseteq A$.*

Proof If $S = s_x \times s_a$, then $S_a = s_x$ for each a and thus, $S_a = S_{a'}$ for all $a, a' \in A$.

Vice versa, let us assume that x and a are independent. Let us denote the common set $S_a = S_{a'}$ by s_x. Let us denote by s_a, the set of all possible values $a \in A$, i.e., the set of all $a \in A$ for which $(x, a) \in S$ for some $x \in X$. Let us prove that in this case, $S = s_x \times s_a$.

Indeed, if $(x, a) \in S$, then, by definition of the set s_x, we have $x \in S_a = s_x$, and, by definition of the set s_a, we have $a \in s_a$. Thus, by the definition of the Cartesian product $B \times C$ as the set of all pairs (b, c) of all pairs of elements $b \in B$ and $c \in C$, we have $(x, a) \in s_x \times s_a$.

Vice versa, let $(x, a) \in s_x \times s_a$, i.e., let $x \in s_x$ and $a \in s_a$. By definition of the set s_x, we have $S_a = s_x$, thus $x \in S_a$. By definition of the set S_a, this means that $(x, a) \in S$. The proposition is proven.

As a corollary, we can conclude that the independence relation is symmetric – similarly to the probabilistic case.

Corollary *Variables x and a are independent if and only if a and x are independent.*

Proof Indeed, both case are equivalent to the condition that the set S is a Cartesian product.

What can we now conclude about the dependence between A, X, and Y. Since we assumed that a and x are independent, we can conclude that the set of possible values of the pair (x, a) is the Cartesian product $X \times A$. For each such pair, the value of y is equal to $y = f(x, a)$. Thus, the set Y is equal to the range of $f(x, a)$ when $x \in X$ and $a \in A$.

The resulting solutions to interval systems of equations. So, we look for sets A for which

$$Y = f(X, A) \stackrel{\text{def}}{=} \{ f(x, a) : x \in X \text{ and } a \in A \}.$$

It is reasonable call the set A satisfying this property an *equality-type solution* to the interval system of equations.

Such solutions for the interval system of equations $y = f(x, a)$, in which we want the interval versions Y and $f(X, A)$ of both sides of the equation to be exactly equal, are known as *algebraic* or, alternatively, *formal* solutions; see, e.g., [11, 12, 45, 62, 83, 94–96].

4.1.3 What if the Interval System of Equations Does not Have an Algebraic (Equality-Type) Solution: A Justification for Enhanced-Zero Solutions

But what if an equality-type solution is impossible: analysis of the problem. The description in the previous section seems to make sense, but sometimes, the corresponding problem has no solutions. For example, in the simplest case when $m = n = k = 1$ and $f(x, a) = x + a$, if we have $Y = [-1, 1]$ and $X = [-2, 2]$, then clearly the corresponding equation $Y = X + A$ does not have a solution: no matter what set A we take the width of the resulting interval $X + A$ is always larger than or equal to the width $w(X) = 4$ of the interval X and thus, cannot be equal to $w(Y) = 2$. What shall we do in this case? How can we then find the desired set A?

Of course, this would not happen if we had the *actual* ranges X and Y, but in reality, we only have estimates for these ranges. So, the fact that we cannot find A means something is wrong with these estimates.

How are ranges X and Y estimated in the first place? To find out what can be wrong, let us recall how the ranges can be obtained from the experiments. For example, in the 1-D case, we perform several measurements of the quantity x_1 in different situations. Based on the corresponding measurement results $x_1^{(c)}$, we conclude that the interval of possible values must include the set $[\underline{x}_1^{\approx}, \overline{x}_1^{\approx}]$, where $\underline{x}_1^{\approx} \stackrel{\text{def}}{=} \min_c x_1^{(c)}$ and $\overline{x}_1^{\approx} \stackrel{\text{def}}{=} \max_c x_1^{(c)}$. Of course, we can also have some values outside this interval – e.g., for a uniform distribution on an interval $[0, 1]$, the interval formed by the smallest and the largest of the C random numbers is slightly narrower than $[0, 1]$; the fewer measurement we take, the narrower this interval.

So, to estimate the actual range, we *inflate* the interval $[\underline{x}_1^{\approx}, \overline{x}_1^{\approx}]$. In these terms, the fact that we have a mismatch between X and Y means that one of these intervals was not inflated enough.

The values x correspond to easier-to-measure quantities, for which we can make a large number of measurements and thus, even without inflation, get pretty accurate estimates of the actual range X. On the other hand, the values y are difficult to measure; for these values, we do not have as many measurement results and thus, there is a need for inflation.

From this viewpoint, we can safely assume that the range for X is reasonably accurate, but the range of Y needs inflation.

So how do we find A? In view of the above analysis, if there is no set A for which $Y = f(X, A)$, the proper solution is to inflate each components of the set Y so that the system becomes solvable.

To make this idea precise, let us formalize what is an inflation.

What is an inflation: analysis of the problem. We want to define a mapping I that transforms each non-degenerate interval $\mathbf{x} = [\underline{x}, \overline{x}]$ into a wider interval

$$I(\mathbf{x}) \supset \mathbf{x}.$$

What are the natural properties of this transformation? The numerical value x of the corresponding quantity depends on the choice of the measuring unit, on the choice of the starting point, and – sometimes – on the choice of direction.

- For example, we can measure temperature t_C in Celsius, but we can also use a different measuring unit and a different starting point and get temperatures in Fahrenheit $t_F = 1.8 \cdot t_C + 32$.
- We can use the usual convention and consider the usual signs of the electric charge, but we could also use the opposite signs – then an electron would be a positive electric charge.

It is reasonable to require that the result of the inflation transformation does not change if we simply change the measuring units or change the starting point or change the sign:

- Changing the starting point leads to a new interval $[\underline{x}, \overline{x}] + x_0 = [\underline{x} + x_0, \overline{x} + x_0]$ for some x_0.
- Changing the measuring unit leads to $\lambda \cdot [\underline{x}, \overline{x}] = [\lambda \cdot \underline{x}, \overline{x}]$ for some $\lambda > 0$.
- Changing the sign leads to $-[\underline{x}, \overline{x}] = [-\overline{x}, -\underline{x}]$.

Thus, we arrive at the following definition.

Definition 4.1.2 By an *inflation operation* (or simply *inflation*, for short), we mean a mapping that maps each non-degenerate interval $\mathbf{x} = [\underline{x}, \overline{x}]$ into a wider interval $I(\mathbf{x}) \supset \mathbf{x}$ so that:

- for every x_0, we have $I(\mathbf{x} + x_0) = I(\mathbf{x}) + x_0$;
- for every $\lambda > 0$, we have $I(\lambda \cdot \mathbf{x}) = \lambda \cdot I(\mathbf{x})$; and
- we have $I(-\mathbf{x}) = -I(\mathbf{x})$.

Proposition 4.1.2 *Every inflation operation has the form*

$$[\tilde{x} - \Delta, \tilde{x} + \Delta] \rightarrow [\tilde{x} - \alpha \cdot \Delta, \tilde{x} + \alpha \cdot \Delta]$$

for some $\alpha > 1$.

Comment A similar result was proven in [37].

Proof It is easy to see that the above operation satisfies all the properties of an inflation. Let us prove that, vice versa, every inflation has this form.

Indeed, for intervals $\mathbf{x}$ of type $[-\Delta, \Delta]$, we have $-\mathbf{x} = \mathbf{x}$, thus $I(\mathbf{x}) = I(-\mathbf{x})$. On the other hand, due to the third property of an inflation, we should have $I(-\mathbf{x}) = -I(\mathbf{x})$. Thus, for the interval $[\underline{v}, \overline{v}] \stackrel{\text{def}}{=} I(\mathbf{x})$, we should have $-[\underline{v}, \overline{v}] = [-\overline{v}, -\underline{v}] = [\underline{v}, \overline{v}]$ and thus, $\underline{v} = -\overline{v}$. So, we have $I([-\Delta, \Delta]) = [-\Delta'(\Delta), \Delta'(\Delta)]$ for some Δ' depending on Δ. Since we should have $[-\Delta, \Delta] \subset I([-\Delta, \Delta])$, we must have

$$\Delta'(\Delta) > \Delta.$$

Let us denote $\Delta'(1)$ by α. Then, $\alpha > 1$ and $I([-1, 1]) = [-\alpha, \alpha]$. By applying the second property of the inflation, with $\lambda = \Delta$, we can then conclude that $I([-\Delta, \Delta]) = [-\alpha \cdot \Delta, \alpha \cdot \Delta]$. By applying the first property of the inflation operation, with $x_0 = \tilde{x}$, we get the desired equality

$$I([\tilde{x} - \Delta, \tilde{x} + \Delta]) = [\tilde{x} - \alpha \cdot \Delta, \tilde{x} + \alpha \cdot \Delta].$$

The proposition is proven.

So how do we find A? We want to make sure that $f(X, A)$ is equal to the result of a proper inflation of Y.

How can we tell that an interval Y' is the result of a proper inflation of Y? One can check that this is equivalent to the fact that the difference $Y' - Y$ is a symmetric interval containing 0; such intervals are known as *extended zeros* [92, 93].

Thus, if we cannot find the set A for which $Y = f(X, A)$, we should look for the set A for which the difference $f(X, A) - Y$ is an extended zero.

Historical comment. This idea was first described in [92, 93]; in this section, we provide a new theoretical justification of this idea.

Multi-D case. What if we have several variables, i.e., $m > 1$? In this case, we may have different inflations for different components Y_i of the set Y, so we should look for the set A for which, for all i, the corresponding difference $f_i(X, A) - Y_i$ is an extended zero.

4.2 Case of Interval Uncertainty: Explaining the Need for Non-realistic Monte-Carlo Simulations

4.2.1 Problem: The Existing Monte-Carlo Method for Interval Uncertainty Is not Realistic

Monte-Carlo method for the case of probabilistic uncertainty is realistic. The probabilistic Monte-Carlo method is *realistic* in the following sense:

- we know that each measurement error Δx_i is distributed according to the probability distribution $\rho_i(\Delta x_i)$, and
- this is exactly how we simulate the measurement errors: to simulate each value $\Delta_i^{(k)}$, we use the exact same distribution $\rho_i(\Delta x_i)$.

In contrast, the Monte-Carlo method for the case of interval uncertainty is not realistic. In the case of uncertainty, all we know is that the measurement errors are always located within the corresponding interval $[-\Delta_i, \Delta_i]$. We do not know how frequently measurement errors will be observed in different parts of this interval. In other words, we do not know the probability distribution of the measurement errors – we only know that this (unknown) probability distribution is located on the interval $[-\Delta_i, \Delta_i]$ *with probability 1.*

With this in mind, a *realistic* Monte-Carlo simulation would mean that for simulating the values $\Delta_i^{(k)}$, we select a probability distribution is located on the corresponding interval $[-\Delta_i, \Delta_i]$ with probability 1. Instead, the existing Monte-Carlo method for interval uncertainty uses Cauchy distribution – and it is known that for this distribution, for any interval, there is a non-zero probability to be outside this interval, and thus, the *probability* to be inside the interval $[-\Delta_i, \Delta_i]$ *is smaller than 1.*

A natural question. A natural question is:

- is this a limitation of the existing method, and an alternative realistic Monte-Carlo method is possible for the case of interval uncertainty,
- or this is a limitation of the problem, and no realistic Monte-Carlo method is possible for interval uncertainty.

What we do in this section. In the two following subsections, we prove that the non-realistic character of the existing Monte-Carlo method for interval uncertainty is a limitation of the problem. In other words, we prove that no realistic Monte-Carlo is possible for the case of interval uncertainty.

In an additional section, we explain why Cauchy distribution should be used.

The results from this section first appeared in [72].

4.2.2 Proof That Realistic Interval Monte-Carlo Techniques Are not Possible: Case of Independent Variables

To prove the desired result, it is sufficient to consider a simple case. To prove the desired impossibility result – that no realistic Monte-Carlo algorithm is possible that would *always* compute the desired range $\mathbf{y}$ – it is sufficient to prove that we cannot get the correct estimate for *one* specific function $f(x_1, \ldots, x_n)$.

As such a function, let us consider the simple function $f(x_1, \ldots, x_n) = x_1 + \cdots + x_n$. In this case, all the partial derivatives are equal to 1, i.e., $c_1 = \cdots = c_n = 1$ and thus,

$$\Delta y = \Delta x_1 + \cdots + \Delta x_n. \tag{4.2.1}$$

If we assume that each variables Δx_i takes value from the interval $[-\delta, \delta]$, then the range of possible values of the sum is $[-\Delta, \Delta]$, where $\Delta = n \cdot \delta$.

Analysis of the problem. Under Monte-Carlo simulations, we have

$$\Delta y^{(k)} = \Delta x_1^{(k)} + \cdots + \Delta x_n^{(k)}. \tag{4.2.2}$$

We assumed that the probability distributions corresponding to all i are independent.

Since the original problem is symmetric with respect to permutations, the corresponding distribution is also symmetric, so all $\Delta_i^{(k)}$ are identically distributed. Thus, the value Δy is the sum of several (n) independent identically distributed random variables.

It is known that due to the Central Limit Theorem (see, e.g., [97]), when n increases, the distribution of the sum tends to Gaussian. So, for large n, this distribution is close to Gaussian.

The Gaussian distribution is uniquely determined by its mean μ and variance $V = \sigma^2$. The mean of the sum is equal to the sum of the means, so $\mu = n \cdot \mu_0$, where μ_0 is the mean of the distribution used to simulate each Δx_i. For independent random variables, the variance of the sum is equal to the sum of the variances, so

$V = n \cdot V_0$, where V_0 is the variance of the distribution used to simulate each Δx_i. Thus, $\sigma = \sqrt{V} = \sqrt{V_0} \cdot \sqrt{n}$.

It is well known that for a normal distribution, with very high confidence, all the values are contained in a k-sigma interval $[\mu - k \cdot \sigma, \mu + k \cdot \sigma]$:

- with probability $\approx 99.9\%$, the value will be in 3-sigma interval,
- with probability $\approx 1 - 10^{-8}$, the value will be in the 6-sigma interval, etc.

Thus, with high confidence, all the values obtained from simulation are contained in the interval $[\mu - k \cdot \sigma, \mu + k \cdot \sigma]$ of width $2k \cdot \sigma = 2k \cdot \sqrt{V_0} \cdot \sqrt{n}$.

For large n, this interval has the size const $\cdot \sqrt{n}$. On the other hand, we want the range $[-\Delta, \Delta]$ whose width is $2\Delta = 2\delta \cdot n$. So, when n is large, the simulated values occupy a part of the desired interval that tends to 0:

$$\frac{2k \cdot \sqrt{V_0} \cdot \sqrt{n}}{2\delta \cdot n} = \frac{\text{const}}{\sqrt{n}} \to 0. \qquad (4.2.3)$$

So, in the independence case, the impossibility is proven.

4.2.3   Proof That Realistic Interval Monte-Carlo Techniques Are not Possible: General Case

To prove the desired negative result, it is sufficient to consider a simple case. Similarly to the previous section, to prove the impossibility result in the *general* case, it is also sufficient to prove the impossibility for *some* of the functions.

In this proof, we will consider functions

$$f(x_1, \ldots, x_n) = s_1 \cdot x_1 + \cdots + s_n \cdot x_n, \qquad (4.2.4)$$

where $s_i \in \{-1, 1\}$.

For each of these functions,

$$\Delta y = s_1 \cdot \Delta x_1 + \cdots + s_n \cdot \Delta x_n, \qquad (4.2.5)$$

so we have $c_i = s_i$. Similarly to the previous section, we assume that each of the unknowns Δx_i takes value from the interval $[-\delta, \delta]$, for some known value $\delta > 0$.

For each of these functions, $|c_i| = |s_i| = 1$, so the desired range is the same for all these functions and is equal to $[-\Delta, \Delta]$, where

$$\Delta = \sum_{i=1}^{n} |c_i| \cdot \Delta_i = n \cdot \delta. \qquad (4.2.6)$$

Towards a precise formulation of the problem. Suppose that we want to find the range $[-\Delta, \Delta]$ with some relative accuracy ε. To get the range from simulations, we need to make sure that some of the simulated results are ε-close to Δ, i.e., that

$$\left| \sum_{i=1}^{n} s_i \cdot \Delta x_i^{(k)} - n \cdot \delta \right| \leq \varepsilon \cdot n \cdot \delta, \tag{4.2.7}$$

or, equivalently,

$$n \cdot \delta \cdot (1 - \varepsilon) \leq \sum_{i=1}^{n} s_i \cdot \Delta x_i^{(k)} \leq n \cdot \delta \cdot (1 + \varepsilon). \tag{4.2.8}$$

We are interested in realistic Monte-Carlo simulations, for which $|\Delta_i^{(k)}| \leq \delta$ for all i. Thus, we always have

$$\sum_{i=1}^{n} s_i \cdot \Delta x_i^{(k)} \leq n \cdot \delta < n \cdot \delta \cdot (1 + \varepsilon). \tag{4.2.9}$$

So, the right-hand inequality is always satisfied, and it is thus sufficient to make sure that we have

$$\sum_{i=1}^{n} s_i \cdot \Delta x_i^{(k)} > n \cdot \delta \cdot (1 - \varepsilon) \tag{4.2.10}$$

for some simulation k.

For this inequality to be true with some certainty, we need to make sure that the probability of this inequality exceed some constant $p > 0$. Then, if we run $1/p$ simulations, then with high probability, the inequality will be satisfied for at least one of these simulations. Thus, we arrive at the following condition.

Definition 4.2.1 Let $\varepsilon > 0$, $\delta > 0$, and $p \in (0, 1)$. We say that a probability distribution on the set of all vectors

$$(\Delta_1 \ldots, \Delta x_n) \in [-\delta, \delta] \times \cdots \times [-\delta, \delta] \tag{4.2.11}$$

is a (p, ε)-realistic Monte-Carlo estimation of interval uncertainty if for every set of values $s_i \in \{-1, 1\}$, we have

$$\text{Prob}(s_1 \cdot \Delta x_1 + \cdots + s_n \cdot \Delta x_n \geq n \cdot \delta \cdot (1 - \varepsilon)) \geq p. \tag{4.2.12}$$

Proposition 4.2.1 *Let $\delta > 0$ and $\varepsilon > 0$. If for every n, we have a (p_n, ε)-realistic Monte-Carlo estimation of interval uncertainty, then $p_n \leq \beta \cdot n \cdot c^n$ for some $\beta > 0$ and $c < 1$.*

Comments.

- As we have mentioned, when the probability is equal to p, we need $1/p$ simulations to get the desired estimates. Due to Proposition 4.2.1, to get a realistic Monte-Carlo estimate for the interval uncertainty, we thus need

$$\frac{1}{p_n} \sim \frac{c^{-n}}{\beta \cdot n} \qquad (4.2.13)$$

simulations. For large n, we have

$$\frac{c^{-n}}{\beta \cdot n} \gg n + 1. \qquad (4.2.14)$$

Thus, the above results shows that realistic Monte-Carlo simulations require even more computational time than numerical differentiation. This defeats the main purpose for using Monte-Carlo techniques, which is – for our problem – to decrease the computation time.

- It is worth mentioning that if we allow p_n to be exponentially decreasing, then a realistic Monte-Carlo estimation of interval uncertainty is possible: e.g., we can take Δx_i to be independent and equal to δ or to $-\delta$ with equal probability 0.5. In this case, with probability 2^{-n}, we get the values $\Delta x_i = s_i \cdot \delta$ for which

$$\sum_{i=1}^{n} s_i \cdot \Delta x_i = \sum_{i=1}^{n} \delta = n \cdot \delta > n \cdot \delta \cdot (1 - \varepsilon). \qquad (4.2.15)$$

Thus, for this probability distribution, for each combination of signs s_i, we have

$$\text{Prob}(s_1 \cdot \Delta x_1 + \cdots + s_n \cdot \Delta x_n \geq n \cdot \delta \cdot (1 - \varepsilon)) = p_n = 2^{-n}. \qquad (4.2.16)$$

Proof of Proposition 4.2.1 Let us pick some $\alpha \in (0, 1)$. Let us denote, by m, the number of indices i or which $s_i \cdot \Delta x_i > \alpha \cdot \delta$. Then, if we have

$$s_1 \cdot \Delta x_1 + \cdots + s_n \cdot \Delta x_n \geq n \cdot \delta \cdot (1 - \varepsilon), \qquad (4.2.17)$$

then for $n - m$ indices, we have $s_i \cdot \Delta x_i \leq \alpha \cdot \delta$ and for the other m indices, we have $s_i \cdot \Delta x_i \leq \delta$. Thus,

$$n \cdot \delta \cdot (1 - \varepsilon) \leq \sum_{i=1}^{n} s_i \cdot \Delta x_i \leq m \cdot \delta + (n - m) \cdot \alpha \cdot \delta. \qquad (4.2.18)$$

Dividing both sides of this inequality by δ, we get

$$n \cdot (1 - \varepsilon) \leq m + (n - m) \cdot \alpha, \qquad (4.2.19)$$

hence $n \cdot (1 - \alpha - \varepsilon) \le m \cdot (1 - \alpha)$ and thus,

$$m \ge n \cdot \frac{1 - \alpha - \varepsilon}{1 - \alpha}. \tag{4.2.20}$$

So, we have at least

$$n \cdot \frac{1 - \alpha - \varepsilon}{1 - \alpha} \tag{4.2.21}$$

indices for which Δx_i has the same sign as s_i (and for which $|\Delta x_i| > \alpha \cdot \delta$). This means that for the vector corresponding to a tuple $(s_1, \ldots, s_n)$, at most

$$n \cdot \frac{\varepsilon}{1 - \alpha - \varepsilon} \tag{4.2.22}$$

indices have a different sign than s_i.

It is, in principle, possible that the same tuple $(\Delta x_1, \ldots, \Delta x_n)$ can serve two different tuples $s = (s_1, \ldots, s_n)$ and $s' = (s'_1, \ldots, s'_n)$. However, in this case:

- going from s_i to $\text{sign}(\Delta x_i)$ changes at most $n \cdot \dfrac{\varepsilon}{1 - \alpha - \varepsilon}$ signs, and
- going from $\text{sign}(\Delta x_i)$ to s'_i also changes at most $n \cdot \dfrac{\varepsilon}{1 - \alpha - \varepsilon}$ signs.

Thus, between the tuples s and s', at most $2 \cdot \dfrac{\varepsilon}{1 - \alpha - \varepsilon}$ signs are different. In other words, for the Hamming distance

$$d(s, s') \stackrel{\text{def}}{=} \#\{i : s_i \ne s'_i\}, \tag{4.2.23}$$

we have

$$d(s, s') \le 2 \cdot n \cdot \frac{\varepsilon}{1 - \alpha - \varepsilon}. \tag{4.2.24}$$

Thus, if

$$d(s, s') > 2 \cdot n \cdot \frac{\varepsilon}{1 - \alpha - \varepsilon}, \tag{4.2.25}$$

then no tuples $(\Delta x_1, \ldots, \Delta x_n)$ can serve both sign tuples s and s'. In this case, the corresponding sets of tuples for which

$$s_1 \cdot \Delta x_1 + \cdots + s_n \cdot \Delta x_n \ge n \cdot \delta \cdot (1 - \varepsilon) \tag{4.2.26}$$

and

$$s'_1 \cdot \Delta x_1 + \cdots + s'_n \cdot \Delta x_n \ge n \cdot \delta \cdot (1 - \varepsilon) \tag{4.2.27}$$

do not intersect. Hence, the probability that the randomly selected tuple belongs to one of these sets is equal to the sum of the corresponding probabilities. Since each of the probabilities is greater than or equal to p, the resulting probability is equal to $2p$.

If we have M sign tuples $s^{(1)}, \ldots, s^{(M)}$ for which

$$d(s^{(i)}, s^{(j)}) > 2 \cdot \frac{\varepsilon}{1 - \alpha - \varepsilon} \tag{4.2.28}$$

for all $i \neq j$, then similarly, the probability that the tuple $(\Delta x_1, \ldots, \Delta x_n)$ serves one of these sign tuples is greater than or equal to $M \cdot p$. On the other hand, this probability is ≤ 1, so we conclude that $M \cdot p \leq 1$ and $p \leq \dfrac{1}{M}$.

So, to prove that p_n is exponentially decreasing, it is sufficient to find the sign tuples whose number M is exponentially increasing.

Let us denote $\beta \stackrel{\text{def}}{=} \dfrac{\varepsilon}{1 - \alpha - \varepsilon}$. Then, for each sign tuple s, the number t of all sign tuples s' for which $d(s, s') \leq \beta \cdot n$ is equal to the sum of:

- the number of tuples $\dbinom{n}{0}$ that differ from s in 0 places,

- the number of tuples $\dbinom{n}{1}$ that differ from s in 1 place, …,

- the number of tuples $\dbinom{n}{\beta \cdot n}$ that differ from s in $\beta \cdot n$ places,

i.e.,

$$t = \binom{n}{0} + \binom{n}{1} + \cdots + \binom{n}{n \cdot \beta}. \tag{4.2.29}$$

When $\beta < 0.5$ and $\beta \cdot n < \dfrac{n}{2}$, the number of combinations $\dbinom{n}{k}$ increases with k, so $t \leq \beta \cdot n \cdot \dbinom{n}{\beta \cdot n}$. Here,

$$\binom{a}{b} = \frac{a!}{b! \cdot (a - b)!}. \tag{4.2.30}$$

Asymptotically,

$$n! \sim \left(\frac{n}{e}\right)^n, \tag{4.2.31}$$

so

$$t \leq \beta \cdot n \cdot \frac{\left(\dfrac{n}{e}\right)^n}{\left(\dfrac{\beta \cdot n}{e}\right)^{\beta \cdot n} \cdot \left(\dfrac{(1 - \beta) \cdot n}{e}\right)^{(1 - \beta) \cdot n}}. \tag{4.2.32}$$

One can see that the term n^n in the numerator cancels with the term $n^{\beta \cdot n} \cdot n^{(1-\beta) \cdot n} = n^n$ in the denominator. Similarly, the terms e^n and $e^{\beta \cdot n} \cdot e^{(1-\beta) \cdot n} = e^n$ cancel each other, so we conclude that

$$t \le \beta \cdot n \cdot \left(\frac{1}{\beta^\beta \cdot (1-\beta)^{1-\beta}} \right)^n. \tag{4.2.33}$$

Here,

$$\gamma \overset{\text{def}}{=} \frac{1}{\beta^\beta \cdot (1-\beta)^{1-\beta}} = \exp(S), \tag{4.2.34}$$

where

$$S \overset{\text{def}}{=} -\beta \cdot \ln(\beta) - (1-\beta) \cdot \ln(1-\beta) \tag{4.2.35}$$

is Shannon's entropy. It is well known (and easy to check by differentiation) that its largest possible values is attained when $\beta = 0.5$, in which case $S = \ln(2)$ and $\gamma = \exp(S) = 2$. When $\beta < 0.5$, we have $S < \ln(2)$, thus, $\gamma < 2$, and $t \le \beta \cdot n \cdot \gamma^n$ for some $\gamma < 2$.

Let us now construct the desired collection of sign tuples $s^{(1)}, \ldots, s^{(M)}$.

- We start with some sign tuple $s^{(1)}$, e.g., $s^{(1)} = (1, \ldots, 1)$.
- Then, we dismiss $t \le \gamma^n$ tuples which are $\le \beta$-close to s, and select one of the remaining tuples as $s^{(2)}$.
- We then dismiss $t \le \gamma^n$ tuples which are $\le \beta$-close to $s^{(2)}$. Among the remaining tuples, we select the tuple $s^{(3)}$, etc.

Once we have selected M tuples, we have thus dismissed $t \cdot M \le \beta \cdot n \cdot \gamma^n \cdot M$ sign tuples. So, as long as this number is smaller than the overall number 2^n of sign tuples, we can continue selecting.

This procedure ends when we have selected M tuples for which $\beta \cdot n \cdot \gamma^n \cdot M \ge 2^n$. Thus, we have selected

$$M \ge \left(\frac{2}{\gamma} \right)^n \cdot \frac{1}{\beta \cdot n} \tag{4.2.36}$$

tuples. So, we have indeed selected exponentially many tuples.

Hence,

$$p_n \le \frac{1}{M} \le \beta \cdot n \cdot \left(\frac{\gamma}{2} \right)^n, \tag{4.2.37}$$

i.e.,

$$p_n \le \beta \cdot n \cdot c^n, \tag{4.2.38}$$

where

$$c \overset{\text{def}}{=} \frac{\gamma}{2} < 1. \tag{4.2.39}$$

So, the probability p_n is indeed exponentially decreasing. The proposition is proven.

4.2.4 Why Cauchy Distribution

Formulation of the problem. We want to find a family of probability distributions with the following property:

- when we have several independent variables $X_1, \ldots, X_n$ distributed according to a distribution with this family with parameters $\Delta_1, \ldots, \Delta_n$,
- then each linear combination $Y = c_1 \cdot X_1 + \cdots + c_n \cdot X_n$ has the same distribution as $\Delta \cdot X$, where X corresponds to parameter 1, and $\Delta = \sum_{i=1}^{n} |c_i| \cdot \Delta_i$.

In particular, for the case when $\Delta_1 = \cdots = \Delta_n = 1$, the problem becomes even easier to describe, since then, we only need to find *one* probability distribution: corresponding to the value 1. In this case, the desired property of this probability distribution is as follows:

- if we have n independent identically distributed random variables $X_1, \ldots, X_n$,
- then each linear combination $Y = c_1 \cdot X_1 + \cdots + c_n \cdot X_n$ has the same distribution as $\Delta \cdot X_i$, where $\Delta = \sum_{i=1}^{n} |c_i|$.

Let us describe all probability distributions that satisfy this property.

Analysis of the problem. First, we observe that from the above condition, for $n = 1$ and $c_1 = -1$, we conclude that $-X$ and X should have exactly the same probability distribution, i.e., that the desired probability distribution be symmetric with respect to 0 (even).

A usual way to describe a probability distribution is to use a probability density function $\rho(x)$, but often, it is more convenient to use its Fourier transform, i.e., in probabilistic terms, the *characteristic function* $\chi_X(\omega) \overset{\text{def}}{=} E[\exp(i \cdot \omega \cdot X)]$, where $E[.]$ indicates the expected value of the corresponding quantity and $i \overset{\text{def}}{=} \sqrt{-1}$.

The advantage of using a characteristic function is that for the sum $S = X_1 + X_2$ of two independent variables $X_1 + X_2$, we have

$$\chi_S(\omega) = E[\exp(i \cdot \omega \cdot S)] = E[\exp(i \cdot \omega \cdot (X_1 + X_2))] = E[\exp(i \cdot \omega \cdot X_1 + i \cdot \omega \cdot X_2)] =$$

$$E[\exp(i \cdot \omega \cdot X_1) \cdot \exp(i \cdot \omega \cdot X_2)]. \tag{4.2.40}$$

Since X_1 and X_2 are independent, the variables $\exp(i \cdot \omega \cdot X_1)$ and $\exp(i \cdot \omega \cdot X_2)$ are also independent, and thus,

$$\chi_S(\omega) = E[\exp(i \cdot \omega \cdot X_1) \cdot \exp(i \cdot \omega \cdot X_2)] = E[\exp(i \cdot \omega \cdot X_1)] \cdot E[\exp(i \cdot \omega \cdot X_2)] =$$

$$\chi_{X_1}(\omega) \cdot \chi_{X_2}(\omega). \tag{4.2.41}$$

Similarly, for a linear combination $Y = \sum_{i=1}^{n} c_i \cdot X_i$, we have

$$\chi_Y(\omega) = E[\exp(i \cdot \omega \cdot Y)] = E\left[\exp\left(i \cdot \omega \cdot \sum_{i=1}^{n} c_i \cdot X_i\right)\right] =$$

$$E\left[\exp\left(\sum_{i=1}^{n} i \cdot \omega \cdot c_i \cdot X_i\right)\right] =$$

$$E\left[\prod_{i=1}^{n} \exp\left(i \cdot \omega \cdot c_i \cdot X_i\right)\right] = \prod_{i=1}^{n} E[\exp(i \cdot (\omega \cdot c_i) \cdot X_i] = \prod_{i=1}^{n} \chi_X(\omega \cdot c_i).$$

$$(4.2.42)$$

The desired property is that the linear combination Y should have the same distribution as $\Delta \cdot X$. Thus, the characteristic function $\chi_Y(\omega)$ should be equal to the characteristic function of $\Delta \cdot X$, i.e., to

$$\chi_{\Delta \cdot X}(\omega) = E[\exp(i \cdot \omega \cdot (\Delta \cdot X))] = E[\exp(i \cdot (\omega \cdot \Delta) \cdot X)] = \chi_X(\omega \cdot \Delta).$$

$$(4.2.43)$$

By comparing expressions (4.2.42) and (4.2.43), we conclude that for all possible combinations $c_1, \ldots, c_n$, the desired characteristic function $\chi_X(\omega)$ should satisfy the equality

$$\chi_X(c_1 \cdot \omega) \cdot \cdots \cdot \chi_X(c_n \cdot \omega) = \chi_X((|c_1| + \cdots + |c_n|) \cdot \omega). \qquad (4.2.44)$$

In particular, for $n = 1$, $c_1 = -1$, we get $\chi_X(-\omega) = \chi_X(\omega)$, so $\chi_X(\omega)$ should be an even function. For $n = 2$, $c_1 > 0$, $c_2 > 0$, and $\omega = 1$, we get

$$\chi_X(c_1 + c_2) = \chi_X(c_1) \cdot \chi_X(c_2). \qquad (4.2.45)$$

The characteristic function should be measurable, and it is known that the only measurable function with the property (4.2.45) has the form $\chi_X(\omega) = \exp(-k \cdot \omega)$ for some k; see, e.g., [1]. Due to evenness, for a general ω, we get $\chi_X(\omega) = \exp(-k \cdot |\omega|)$. By applying the inverse Fourier transform, we conclude that X is Cauchy distributed.

Conclusion. The only distribution for which the independent-case Monte Carlo simulations lead to correct estimate of the interval uncertainty is the Cauchy distribution.

4.3   Case of Probabilistic Uncertainty: Why Mixture of Probability Distributions?

If we have two random variables ξ_1 and ξ_2, then we can form their *mixture* if we take ξ_1 with some probability w and ξ_2 with the remaining probability $1 - w$. The probability density function (pdf) $\rho(x)$ of the mixture is a convex combination of the pdfs of the original variables: $\rho(x) = w \cdot \rho_1(x) + (1 - w) \cdot \rho_2(x)$. A natural question is:

can we use other functions $f(\rho_1, \rho_2)$ to combine the pdfs, i.e., to produce a new pdf $\rho(x) = f(\rho_1(x), \rho_2(x))$? In this section, we prove that the only combination operations that always lead to a pdf are the operations

$$f(\rho_1, \rho_2) = w \cdot \rho_1 + (1 - w) \cdot \rho_2$$

corresponding to mixture.

Results from this section first appeared in [75].

4.3.1   Formulation of the Problem

What is mixture. If we have two random variables ξ_1 and ξ_2, then, for each probability $w \in [0, 1]$, we can form a *mixture* ξ of these variables by selecting ξ_1 with probability w and ξ_2 with the remaining probability $1 - w$; see, e.g., [97].

In particular, if we know the probability density function (pdf) $\rho_1(x)$ corresponding to the first random variable and the probability density function $\rho_2(x)$ corresponding to the second random variable, then the probability density function $\rho(x)$ corresponding to their mixture has the usual form

$$\rho(x) = w \cdot \rho_1(x) + (1 - w) \cdot \rho_2(x). \tag{4.3.1}$$

A natural question. A natural question is: are there other combination operations $f(\rho_1, \rho_2)$ that always transform two probability distributions $\rho_1(x)$ and $\rho_2(x)$ into a new probability distribution

$$\rho(x) = f(\rho_1(x), \rho_2(x)). \tag{4.3.2}$$

Our result. Our result is that the only possible transformation (4.3.2) that always generates a probability distribution is the mixture (4.3.1), for which

$$f(\rho_1, \rho_2) = w \cdot \rho_1 + (1 - w) \cdot \rho_2 \tag{4.3.3}$$

for some $w \in [0, 1]$.

4.3.2   Main Result of This Section

Definition 4.3.1 We say that a function $f(\rho_1, \rho_2)$ that maps pairs of non-negative real numbers into a non-negative real number is a *probability combination operation* if for every two probability density functions $\rho_1(x)$ and $\rho_2(x)$ defined on the same set X, the function $\rho(x) = f(\rho_1(x), \rho_2(x))$ is also a probability density function, i.e., $\int \rho(x) \, dx = 1$.

Proposition 4.3.1 *A function* $f(\rho_1, \rho_2)$ *is a probability combination operation if and only if it has the form* $f(\rho_1, \rho_2) = w \cdot \rho_1 + (1 - w) \cdot \rho_2$ *for some* $w \in [0, 1]$.

Proof
$1°$. Let us first prove that $f(0, 0) = 1$.
  Indeed, let us take $X = \mathbb{R}$, and the following pdfs:

- $\rho_1(x) = \rho_2(x) = 1$ for $x \in [0, 1]$ and
- $\rho_1(x) = \rho_2(x) = 0$ for all other values x.

Then, the combined function $\rho(x) = f(\rho_1(x), \rho_2(x))$ has the following form:

- $\rho(x) = f(1, 1)$ when $x \in [0, 1]$ and
- $\rho(x) = f(0, 0)$ for $x \notin [0, 1]$.

Let us use the condition $\int \rho(x)\,dx = 1$ to prove that $f(0, 0) = 0$.
  We can prove it by contradiction. If we had $f(0, 0) \neq 0$, i.e., if we had $f(0, 0) > 0$, then we would have

$$\int \rho(x)\,dx = f(1, 1) \cdot 1 + f(0, 0) \cdot \infty = \infty \neq 1.$$

Thus, we should have $f(0, 0) = 0$.
$2°$. Let us now prove that $f(0, \rho_2) = k_2 \cdot \rho_2$ for some $k_2 \geq 0$.
  Let us take the following function $\rho_1(x)$:

- $\rho_1(x) = 1$ for $x \in [-1, 0]$ and
- $\rho_1(x) = 0$ for all other x.

  Let us now pick any number $\rho_2 > 0$ and define the following pdf $\rho_2(x)$:

- $\rho_2(x) = \rho_2$ for $x \in [0, 1/\rho_2]$ and
- $\rho_2(x) = 0$ for all other x.

In this case, the combined function $\rho(x)$ has the following form:

- $\rho(x) = f(1, 0)$ for $x \in [-1, 0]$;
- $\rho(x) = f(0, \rho_2)$ for $x \in [0, 1/\rho_2]$, and
- $\rho(x) = f(0, 0) = 0$ for all other x.

Thus, the condition $\int \rho(x)\,dx = 1$ takes the form

$$f(1, 0) + f(0, \rho_2) \cdot (1/\rho_2) = 1,$$

hence $f(0, \rho_2) \cdot (1/\rho_2) = 1 - f(1, 0)$ and therefore, $f(0, \rho_2) = k_2 \cdot \rho_2$, where we denoted $k_2 \stackrel{\text{def}}{=} 1 - f(1, 0)$.
$3°$. Similarly, we can prove that $f(\rho_1, 0) = k_1 \cdot \rho_1$ for some $k_1 \geq 0$.
$4°$. Let us now prove that for all ρ_1 and ρ_2, we have $f(\rho_1, \rho_2) = k_1 \cdot \rho_1 + k_2 \cdot \rho_2$.
  We already know, from Parts 1, 2 and 3 of this proof, that the desired equality holds when one of the values ρ_i is equal to 0.

Let us now take any values $\rho_1 > 0$ and $\rho_2 > 0$. Let us then pick a positive value $\Delta \leq 1/\max(\rho_1, \rho_2)$ and define the following pdfs. The first pdf $\rho_1(x)$ is defined by the following formulas:

- $\rho_1(x) = \rho_1$ for $x \in [0, \Delta]$,
- $\rho_1(x) = 1$ for $x \in [-(1 - \Delta \cdot \rho_1), 0]$, and
- $\rho_1(x) = 0$ for all other x.

The second pdf $\rho_2(x)$ is defined by the following formula:

- $\rho_2(x) = \rho_2$ for $x \in [0, \Delta]$,
- $\rho_2(x) = 1$ for $x \in [\Delta, \Delta + (1 - \Delta \cdot \rho_2)]$, and
- $\rho_2(x) = 0$ for all other x.

Then, the combined function $\rho(x) = f(\rho_1(x), \rho_2(x))$ has the following form

- $\rho(x) = f(1, 0) = k_1$ for $x \in [-(1 - \Delta \cdot \rho_1), 0]$,
- $\rho(x) = f(\rho_1, \rho_2)$ for $x \in [0, \Delta]$,
- $\rho(x) = f(0, 1) = k_2$ for $x \in [\Delta, \Delta + (1 - \Delta \cdot \rho_2)]$, and
- $\rho(x) = f(0, 0) = 0$ for all other x.

For this combined function $\rho(x)$, the condition that $\int \rho(x)\, dx = 1$ takes the form

$$k_1 \cdot (1 - \Delta \cdot \rho_1) + f(\rho_1, \rho_2) \cdot \Delta + k_2 \cdot (1 - \Delta \cdot \rho_2) = 1. \qquad (4.3.4)$$

Let us now consider a different pair of pdfs, $\rho_1'(x)$ and $\rho_2'(x)$. The first pdf $\rho_1'(x)$ is defined by the following formulas:

- $\rho_1'(x) = 2\rho_1$ for $x \in [0, \Delta/2]$,
- $\rho_1'(x) = 1$ for $x \in [-(1 - \Delta \cdot \rho_1), 0]$, and
- $\rho_1'(x) = 0$ for all other x.

The second pdf $\rho_2'(x)$ is defined by the following formula:

- $\rho_2'(x) = 2\rho_2$ for $x \in [\Delta/2, \Delta]$,
- $\rho_2'(x) = 1$ for $x \in [\Delta, \Delta + (1 - \Delta \cdot \rho_2)]$, and
- $\rho_2'(x) = 0$ for all other x.

Then, the combined function $\rho'(x) = f(\rho_1'(x), \rho_2'(x))$ has the following form

- $\rho'(x) = f(1, 0) = k_1$ for $x \in [-(1 - \Delta \cdot \rho_1), 0]$,
- $\rho'(x) = f(2\rho_1, 0) = k_1 \cdot (2\rho_1)$ for $x \in [0, \Delta/2]$,
- $\rho'(x) = f(0, 2\rho_2) = k_2 \cdot (2\rho_2)$ for $x \in [\Delta/2, \Delta]$,
- $\rho'(x) = f(0, 1) = k_2$ for $x \in [\Delta, \Delta + (1 - \Delta \cdot \rho_2)]$, and
- $\rho'(x) = f(0, 0) = 0$ for all other x.

For this combined function $\rho'(x)$, the condition that $\int \rho'(x)\, dx = 1$ takes the form

$$k_1 \cdot (1 - \Delta \cdot \rho_1) + k_1 \cdot (2\rho_1) \cdot (\Delta/2) + k_2 \cdot (2\rho_2) \cdot (\Delta/2) + k_2 \cdot (1 - \Delta \cdot \rho_2) = 1. \qquad (4.3.5)$$

If we subtract (4.3.5) from (4.3.4) and divide the difference by $\Delta > 0$, then we conclude that $f(\rho_1, \rho_2) - k_1 \cdot \rho_1 - k_2 \cdot \rho_2 = 0$, i.e., exactly what we want to prove in this section.

$5°$. To complete the proof, we need to show that $k_2 = 1 - k_1$, i.e., that $k_1 + k_2 = 1$. Indeed, let us take:

- $\rho_1(x) = \rho_2(x) = 1$ when $x \in [0, 1]$ and
- $\rho_1(x) = \rho_2(x)$ for all other x.

Then, for the combined pdf, we have:

- $\rho(x) = f(\rho_1(x), \rho_2(x)) = k_1 + k_2$ for $x \in [0, 1]$ and
- $\rho(x) = 0$ for all other x.

For this combined function $\rho(x)$, the condition $\int \rho(x)\, dx = 1$ implies that

$$k_1 + k_2 = 1.$$

The proposition is proven.

4.4 Case of Fuzzy Uncertainty: Every Sufficiently Complex Logic Is Multi-valued Already

Usually, fuzzy logic (and multi-valued logics in general) are viewed as drastically different from the usual 2-valued logic. In this section, we show that while on the surface, there indeed seems to be a major difference, a more detailed analysis shows that even in the theories based on the 2-valued logic, there naturally appear constructions which are, in effect, multi-valued, constructions which are very close to fuzzy logic.

Results of this section first appeared in [73].

4.4.1 Formulation of the Problem: Bridging the Gap Between Fuzzy Logic and the Traditional 2-Valued Fuzzy Logic

There seems to be a gap. One of the main ideas behind fuzzy logic (see, e.g., [28, 61, 108]) is that:

- in contrast to the traditional 2-valued logic, in which every statement is either true or false,
- in fuzzy logic, we allow intermediate degrees.

In other words, fuzzy logic is an example of a *multi-valued* logic.

This difference has led to some mutual misunderstanding between researchers in fuzzy logic and researchers in traditional logic:

- on the one hand, many popular articles on fuzzy logic, after explaining the need for intermediate degrees, claim that in the traditional 2-valued logic, it is not possible to represent such degrees – and thus, a drastically new logic is needed;
- on the other hand, many researchers from the area of traditional logic criticize fuzzy logic for introducing – in their opinion – "artificial" intermediate degrees, degrees which are contrary to their belief that eventually, every statement is either true or false.

What we plan to show in this section. In this section, we plan to show that the above mutual criticism is largely based on a misunderstanding.

Yes, in the first approximation, there seems to be a major difference between 2-valued and multi-valued logic. However, if we dig deeper and consider more complex constructions, we will see that in the traditional 2-valued logic there are, in effect, multiple logical values.

The usual way of introducing multiple values in 2-valued logics is based on ideas which are different from the usual fuzzy logic motivations. However, we show that there is a reasonably natural alternative way to introduce multi-valuedness into the traditional 2-valued logic, a way which is very similar to what we do in fuzzy logic.

Thus, we bridge the gap between the fuzzy logic and the traditional 2-valued fuzzy logic.

Why this may be interesting? Definitely not from the practical viewpoint; we take simple techniques of fuzzy logic and we interpret them in a rather complex way.

But, from the theoretical viewpoint, we believe that bridging this gap is important. It helps to tone down the usual criticisms:

- Contrary to the opinion which is widely spread in the fuzzy logic community, it *is* possible to describe intermediate degrees in the traditional 2-valued logic. However, such a representation is complicated. The main advantage of fuzzy techniques is that they provide a simply way of doing this – and simplicity is important for applications.
- On the other hand, contrary to the opinion which is widely spread in the classical logic community, the main ideas of fuzzy logic are not necessarily inconsistent with the 2-valued foundations; moreover, they naturally appear in these foundations if we try to adequately describe expert knowledge.

We hope that, after reading this section, researchers from both communities will better understand each other.

The structure of this section. In Sect. 4.4.2, we describe the usual view of fuzzy logic as a technique which is drastically different from the usual 2-valued logic. In Sect. 4.4.3, we remind the readers that in the 2-valued approach, there is already multi-valuedness – although this multi-valuedness is different from what we consider in fuzzy logic. Finally, in Sect. 4.4.4, we show that the 2-valued-logic-based analysis of expert knowledge naturally leads to a new aspect of multi-valuedness, an aspect which is very similar to the main ideas behind fuzzy logic.

4.4.2   Fuzzy Logic – The Way It Is Typically Viewed as Drastically Different from the Traditional 2-Valued Logic

Usual motivations behind fuzzy logic. Our knowledge of the world is rarely absolutely perfect. As a result, when we make decisions, then, in addition to the well-established facts, we have to rely on the human expertise, i.e., on expert statements about which the experts themselves are not 100% confident.

If we had a perfect knowledge, then, for each possible statement, we would know for sure whether this statement is true or false. Since our knowledge is not perfect, for many statements, we are not 100% sure whether they are true or false.

Main idea behind fuzzy logic. To describe and process such statements, Zadeh proposed special *fuzzy logic* techniques, in which, in addition to "true" and "false", we have intermediate degrees of certainty; see, e.g., [28, 61, 108].

In a nutshell, the main idea behind fuzzy logic is to go:

- from the traditional 2-valued logic, in which every statement is either true or false,
- to a multi-valued logic, in which we have more options to describe our opinion about he truth of different statements.

From this viewpoint, the traditional 2-valued logic and the fuzzy logic are drastically different. Namely, these logics correspond to a different number of possible truth values.

4.4.3   There Is Already Multi-valuedness in the Traditional 2-Valued Fuzzy Logic: Known Results

Source of multi-valuedness: Gödel's theorem. At first glance, the difference does seem drastic. However, let us recall that the above description of the traditional 2-valued logic is based on the idealized case when for every statement S, we know whether this statement is true or false.

This is possible in simple situations, but, as the famous Gödel's theorem shows, such an idealized situation is not possible for sufficiently complex theories; see, e.g., [16, 91]. Namely, Gödel proved that already for arithmetic – i.e., for statements obtained from basic equality and inequality statements about polynomial expressions by adding propositional connectives &, $\lor$, $\neg$, and quantifiers over natural numbers – it is not possible to have a theory T in which for every statement S, either this statement or its negation are derived from this theory (i.e., either $T \vDash S$ or $T \vDash \neg S$).

We have, in effect, at least three different truth values. Due to Gödel's theorem, there exist statements S for which $T \nvDash S$ and $T \nvDash \neg S$. So:

- while, legally speaking, the corresponding logic is 2-valued,
- in reality, such a statement S is neither true nor false – and thus, we have more than 2 possible truth values.

At first glance, it may seem that here, we have a 3-valued logic, with possible truth values "true", "false", and "unknown", but in reality, we may have more, since:

- while different "true" statements are all provably equivalent to each other, and
- all "false" statements are provably equivalent to each other,
- different "unknown" statements are not necessarily provably equivalent to each other.

How many truth values do we actually have: the notion of Lindenbaum–Tarski algebra. To get a more adequate description of this situation, it is reasonable to consider the equivalence relation $\vDash (A \Leftrightarrow B)$ between statements A and B.

Equivalence classes with respect to this relation can be viewed as the actual truth values of the corresponding theory. The set of all such equivalence classes is known as the *Lindenbaum–Tarski algebra*; see, e.g., [16, 91].

But what does this have to do with fuzzy logic? Lindenbaum–Tarski algebra shows that any sufficiently complex logic is, in effect, multi-valued.

However, this multi-valuedness is different from the multi-valuedness of fuzzy logic.

What we do in this section. In the next section, we show that there is another aspect of multi-valuedness of the traditional logic, an aspect of which the usual fuzzy logic is, in effect, a particular case. Thus, we show that the gap between the traditional 2-valued logic and the fuzzy logic is even less drastic.

4.4.4 Application of 2-Valued Logic to Expert Knowledge Naturally Leads to a New Aspect of Multi-valuedness – An Aspect Similar to Fuzzy

Need to consider several theories. In the previous section, we considered the case when we have a single theory T.

Gödel's theorem states that for every given theory T that includes formal arithmetic, there is a statement S that can neither be proven nor disproven in this theory. Since this statement S can neither be proven not disproven based on the axioms of theory T, a natural idea is to consider additional reasonable axioms that we can add to T.

This is what happened, e.g., in geometry, with Euclid's Vth postulate – that for every line ℓ in a plane and for every point P outside this line, there exists only one line ℓ' which passes through P and is parallel to ℓ. Since it turned out that neither this statement nor its negation can be derived from all other (more intuitive) axioms of geometry, a natural solution is to explicitly add this statement as a new axiom. (If we add its negation, we get Lobachevsky geometry – historically the first non-Euclidean geometry; see, e.g., [7].)

Similarly, in set theory, it turns out that the Axiom of Choice and Continuum Hypothesis cannot be derived or rejected based on the other (more intuitive) axioms of set theory; thus, they (or their negations) have to be explicitly added to the original theory; see, e.g., [44].

The new – extended – theory covers more statements that the original theory T.

- However, the same Gödel's theory still applies.
- So for the new theory, there are still statements that can neither be deduced nor rejected based on this new theory.
- Thus, we need to add one more axiom, etc.

As a result:

- instead of a *single* theory,
- it makes sense to consider a *family* of theories $\{T_\alpha\}_\alpha$.

In the above description, we end up with a family which is *linearly ordered* in the sense that for every two theories T_α and T_β, either $T_\alpha \vDash T_\beta$ or $T_\beta \vDash T_\alpha$. However, it is possible that on some stage, different groups of researchers select two different axioms – e.g., a statement and its negation. In this case, we will have two theories which are not derivable from each other – and thus a family of theories which is not linearly ordered.

How is all this applicable to expert knowledge? From the logical viewpoint, processing expert knowledge can also be viewed as a particular case of the above scheme: axioms are the basic logical axioms + all the expert statements statements that we believe to be true.

- We can select only the statements in which experts are 100% sure, and we get one possible theory.
- We can add statements S for which the expert's degree of confidence $d(S)$ exceeds a certain threshold α – and get a different theory, with a larger set of statements

$$\{S : d(S) \geq \alpha\}.$$

- Depending on our selection of the threshold α, we thus get different theories T_α.

So, in fact, we also have a family of theories $\{T_\alpha\}_\alpha$, where different theories T_α correspond to different levels of the certainty threshold α.

Example. For example, if we select $\alpha = 0.7$, then:

- For every object x for which the expert's degree of confidence that x is small is at least 0.7, we consider the statement $S(x)$ ("x is small") to be true.
- For all other objects x, we consider $S(x)$ to be false.

Similarly, we only keep "if-then" rules for which the expert's degree of confidence in these rules is either equal to 0.7 or exceeds 0.7.

Once we have a family of theories, how can we describe the truth of a statement? If we have a single theory T, then for every statement S, we have three possible options:

- either $T \vDash S$, i.e., the statement S is true in the theory T,
- or $T \vDash \neg S$, i.e., the statement S is false in the theory T,
- or $T \nvDash S$ and $T \nvDash \neg S$, i.e., the statement S is undecidable in this theory.

Since, as we have mentioned earlier, a more realistic description of our knowledge means that we have to consider a family of theories $\{T_\alpha\}_\alpha$, it is reasonable to collect this information based on all the theories T_α.

Thus, to describe whether a statement S is true or not, instead of a single yes-no value (as in the case of a single theory), we should consider the values corresponding to all the theories T_α, i.e., equivalently, we should consider the whole set

$$\deg(S) \overset{\text{def}}{=} \{\alpha : T_\alpha \vDash S\}.$$

This set is our degree of belief that the statement S is true – i.e., in effect, the truth value of the statement S.

Logical operations on the new truth values. If a theory T_α implies both S and S', then this theory implies their conjunction $S \& S'$ as well. Thus, the truth value of the conjunction includes the intersection of truth value sets corresponding to S and S':

$$\deg(S \& S') \supseteq \deg(S) \cap \deg(S').$$

Similarly, if a theory T_α implies either S or S', then this theory also implies the disjunction $S \vee S'$. Thus, the truth value of the disjunction includes the union of truth value sets corresponding to S and S':

$$\deg(S \vee S') \supseteq \deg(S) \cup \deg(S').$$

What happens in the simplest case, when the theories are linearly ordered? If the theories T_α are linearly ordered, then, once $T_\alpha \vDash S$ and $T_\beta \vDash T_\alpha$, we also have $T_\beta \vDash S$. Thus, with every T_α, the truth value $\deg(S) = \{\alpha : T_\alpha \vDash S\}$ includes, with each index α, the indices of all the stronger theories – i.e., all the theories T_β for which $T_\beta \vDash T_\alpha$.

In particular, in situations when we have a finite family of theories, each degree is equal to $D_{\alpha_0} \overset{\text{def}}{=} \{\alpha : T_\alpha \vDash T_{\alpha_0}\}$ for some α_0. In terms of the corresponding linear order

$$\alpha \leq \beta \Leftrightarrow T_\alpha \vDash T_\beta,$$

this degree takes the form $D_{\alpha_0} = \{\alpha : \alpha \leq \alpha_0\}$. We can thus view α_0 as the degree of truth of the statement S:

$$\text{Deg}(S) \overset{\text{def}}{=} \alpha_0.$$

In case of expert knowledge, this means that we consider the smallest degree of confidence d for which we can derive the statement S if we allow all the expert's statements whose degree of confidence is at least d.

- If we can derive S by using only statements in which the experts are absolutely sure, then we are very confident in this statement S.
- On the other hand, if, in order to derive the statement S, we need to also consider expert's statement in which the experts are only somewhat confident, then, of course, our degree of confidence in S is much smaller.

These sets D_α are also linearly ordered: one can easily show that

$$D_\alpha \subseteq D_\beta \Leftrightarrow \alpha \leq \beta.$$

In this case:

- the intersection of sets D_α and D_β simply means that we consider the set $D_{\min(\alpha,\beta)}$, and
- the union of sets D_α and D_β simply means that we consider the set $D_{\max(\alpha,\beta)}$.

Thus, the above statements about conjunction and disjunction take the form

$$\text{Deg}(S \mathbin{\&} S') \geq \min(\text{Deg}(S), \text{Deg}(S'));$$

$$\text{Deg}(S \vee S') \geq \max(\text{Deg}(S), \text{Deg}(S')).$$

This is very similar to the usual fuzzy logic. The above formulas are very similar to the formulas of the fuzzy logic corresponding to the most widely used "and"- and "or"- operations: min and max. (The only difference is that we get $\geq$ instead of the equality.)

Thus, fuzzy logic ideas can be indeed naturally obtained in the classical 2-valued environment: namely, they can be interpreted as a particular case of the same general idea as the Lindenbaum–Tarski algebra.

4.5 General Case: Explaining Ubiquity of Linear Models

Linear interpolation is the computationally simplest of all possible interpolation techniques. Interestingly, it works reasonably well in many practical situations, even in situations when the corresponding computational models are rather complex. In this section, we explain this empirical fact by showing that linear interpolation is the only interpolation procedure that satisfies several reasonable properties such as consistency and scale-invariance.

Results from this section first appeared in [76].

4.5.1 Formulation of the Problem

Need for interpolation. In many practical situations, we know that the value of a quantity y is uniquely determined by the value of some other quantity x, but we do not know the exact form of the corresponding dependence $y = f(x)$.

To find this dependence, we measure the values of x and y in different situations. As a result, we get the values $y_i = f(x_i)$ of the unknown function $f(x)$ for several values $x_1, \ldots, x_n$. Based on this information, we would like to predict the value $f(x)$ for all other values x. When x is between the smallest and the largest of the values x_i, this prediction is known as the *interpolation*, for values x smaller than the smallest of x_i or larger than the largest of x_i, this prediction is known as *extrapolation*; see, e.g., [9].

Simplest possible case of interpolation. The simplest possible case of interpolation is when we only know the values $y_1 = f(x_1)$ and $y_2 = f(x_2)$ of the function $f(x)$ at two points $x_1 < x_2$, and we would like to predict the value $f(x)$ at points $x \in (x_1, x_2)$.

In many cases, linear interpolations works well: why? One of the most well-known interpolation techniques is based on the assumption that the function $f(x)$ is linear on the interval $[x_1, x_2]$. Under this assumption, we get the following formula for $f(x)$:

$$f(x) = \frac{x - x_1}{x_2 - x_1} \cdot f(x_2) + \frac{x_2 - x}{x_2 - x_1} \cdot f(x_1).$$

This formula is known as *linear interpolation*.

The usual motivation for linear interpolation is simplicity: linear functions are the easiest to compute, and this explains why we use linear interpolation.

An interesting empirical fact is that in many practical situations, linear interpolation works reasonably well. We know that in computational science, often very complex computations are needed, so we cannot claim that nature prefers simple functions. There should be another reason for the empirical fact that linear interpolation often works well.

What we do. In this section, we show that linear interpolation can indeed be derived from fundamental principles.

4.5.2   Analysis of the Problem: What Are Reasonable Property of an Interpolation

What is interpolation. We want to be able, given values y_1 and y_2 of the unknown function at points x_1 and x_2, and a point $x \in (x_1, x_2)$, to provide an estimate for $f(x)$. In other words, we need a function that, given the values x_1, y_1, x_2, y_2, and x, generates the estimate for $f(x)$. We will denote this function by $I(x_1, y_1, x_2, y_2, x)$.

What are the reasonable properties of this function?

Conservativeness. If both observed values $y_i = f(x_i)$ are smaller than or equal to some threshold value y, it is reasonable to expect that all intermediate values of $f(x)$ should also be smaller than or equal to y. Thus, if $y_1 \leq y$ and $y_2 \leq y$, then we should have $I(x_1, y_1, x_2, y_2, x) \leq y$.

In particular, for $y = \max(y_1, y_2)$, we conclude that

$$I(x_1, y_1, x_2, y_2, x) \leq \max(y_1, y_2).$$

Similarly, if both observed values $y_i = f(x_i)$ are greater than or equal to some threshold value y, it is reasonable to expect that all intermediate values of $f(x)$ should also be greater than or equal to y. Thus, if $y \leq y_1$ and $y \leq y_2$, then we should have $y \leq I(x_1, y_1, x_2, y_2, x)$.

In particular, for $y = \min(y_1, y_2)$, we conclude that

$$\min(y_1, y_2) \leq I(x_1, y_1, x_2, y_2, x)$$

These two requirements can be combined into a single double inequality

$$\min(y_1, y_2) \leq I(x_1, y_1, x_2, y_2, x) \leq \max(y_1, y_2).$$

We will call this property *conservativeness*.

x-scale-invariance. The numerical value of a physical quantity depends on the choice of the measuring unit and on the starting point. If we change the starting point to the one which is b units smaller, then b is added to all the numerical values. Similarly, if we replace a measuring unit by a one which is $a > 0$ times smaller, then all the numerical values are multiplied by a. If we perform both changes, then each original value x is replaced by the new value $x' = a \cdot x + b$.

For example, if we know the temperature x in Celsius, then the temperature x' in Fahrenheit can be obtained as $x' = 1.8 \cdot x + 32$.

It is reasonable to require that the interpolation procedure should not change if we simply change the measuring unit and the starting point – without changing the actual physical quantities. In other words, it is reasonable to require that

$$I(a \cdot x_1 + b, y_1, a \cdot x_2 + b, y_2, a \cdot x + b) = I(x_1, y_1, x_2, y_2, x).$$

y-scale-invariance. Similarly, we can consider different units for y. The interpolation result should not change if we simply change the starting point and the measuring unit. So, if we replace y_1 with $a \cdot y_1 + b$ and y_2 with $a \cdot y_2 + b$, then the result of interpolation should be obtained by a similar transformation from the previous result: $I \rightarrow a \cdot I + b$. Thus, we require that

$$I(x_1, a \cdot y_1 + b, x_2, a \cdot y_2 + b, x) = a \cdot I(x_1, y_1, x_2, y_2, x) + b.$$

Consistency. Let us assume that we have $x_1 \leq x_1' \leq x \leq x_2' \leq x_2$. Then, the value $f(x)$ can be estimated in two different ways:

- we can interpolate directly from the values $y_1 = f(x_1)$ and $y_2 = f(x_2)$, getting $I(x_1, y_1, x_2, y_2, x)$, or
- we can first use interpolation to estimate the values $f(x_1') = I(x_1, y_1, x_2, y_2, x_1')$ and $f(x_2') = I(x_1, y_1, x_2, y_2, x_2')$, and then use these two estimates to estimate $f(x)$ as

$$I(x_1, f(x_1'), x_2, f(x_2'), x) =$$

$$I(x_1', I(x_1, y_1, x_2, y_2, x_1'), x_2', I(x_1, y_1, x_2, y_2, x_2'), x).$$

It is reasonable to require that these two ways lead to the same estimate for $f(x)$:

$$I(x_1, y_1, x_2, y_2, x) = I(x_1', I(x_1, y_1, x_2, y_2, x_1'), x_2', I(x_1, y_1, x_2, y_2, x_2'), x).$$

Continuity. Most physical dependencies are continuous. Thus, when the two value x and x' are close, we expect the estimates for $f(x)$ and $f(x')$ to be also close. Thus, it is reasonable to require that the interpolation function $I(x_1, y_1, x_2, y_2, x)$ is continuous in x – and that for both $i = 1, 2$ the value $I(x_1, y_1, x_2, y_2, x)$ converges to $f(x_i)$ when $x \to x_i$.

Now, we are ready to formulate our main result.

4.5.3 Main Result of This Section

Definition 4.5.1 By an *interpolation function*, we mean a function $I(x_1, y_1, x_2, y_2, x)$ which is defined for all $x_1 < x < x_2$ and which has the following properties:

- conservativeness:

$$\min(y_1, y_2) \le I(x_1, y_1, x_2, y_2, x) \le \max(y_1, y_2)$$

for all $x_i, y_i,$ and x;
- x-scale-invariance: $I(a \cdot x_1 + b, y_1, a \cdot x_2 + b, y_2, a \cdot x + b) = I(x_1, y_1, x_2, y_2, x)$ for all $x_i, y_i, x, a > 0,$ and b;
- y-scale invariance: $I(x_1, a \cdot y_1 + b, x_2, a \cdot y_2 + b, x) = a \cdot I(x_1, y_1, x_2, y_2, x) + b$ for all $x_i, y_i, x, a > 0,$ and b;
- consistency:

$$I(x_1, y_1, x_2, y_2, x) = I(x_1', I(x_1, y_1, x_2, y_2, x_1'), x_2', I(x_1, y_1, x_2, y_2, x_2'), x)$$

for all $x_i, x_i', y_i,$ and x; and
- continuity: the expression $I(x_1, y_1, x_2, y_2, x)$ is a continuous function of x, $I(x_1, y_1, x_2, y_2, x) \to y_1$ when $x \to x_1$ and $I(x_1, y_1, x_2, y_2, x) \to y_2$ when $x \to x_2$.

Proposition 4.5.1 *The only interpolation function satisfying all the properties from Definition 4.5.1 is the linear interpolation*

$$I(x_1, y_1, x_2, y_2, x) = \frac{x - x_1}{x_2 - x_1} \cdot y_2 + \frac{x_2 - x}{x_2 - x_1} \cdot y_1. \tag{4.5.1}$$

Discussion. Thus, we have indeed explained that linear interpolation follows from the fundamental principles – which may explain its practical efficiency.

Proof

$1°$. When $y_1 = y_2$, the conservativeness property implies that $I(x_1, y_1, x_2, y_1, x) = y_1$. Thus, to complete the proof, it is sufficient to consider two remaining cases: when $y_1 < y_2$ and when $y_2 < y_1$.

We will consider the case when $y_1 < y_2$. The case when $y_2 < y_1$ is considered similarly. So, in the following text, without losing generality, we assume that

$$y_1 < y_2.$$

$2°$. When $y_1 < y_2$, then we can get these two values y_1 and y_2 as $y_1 = a \cdot 0 + b$ and $y_2 = a \cdot 1 + b$ for $a = y_2 - y_1$ and $b = y_1$. Thus, the y-scale-invariance implies that

$$I(x_1, y_1, x_2, y_2, x) = (y_2 - y_1) \cdot I(x_1, 0, x_2, 1, x) + y_1. \qquad (4.5.2)$$

If we denote $J(x_1, x_2, x) \overset{\text{def}}{=} I(x_1, 0, x_2, 1, x)$, then we get

$$I(x_1, y_1, x_2, y_2, x) = (y_2 - y_1) \cdot J(x_1, x_2, x) + y_1 =$$

$$J(x_1, x_2, x) \cdot y_2 + (1 - J(x_1, x_2, x)) \cdot y_1. \qquad (4.5.3)$$

$3°$. Since $x_1 < x_2$, we can similarly get these two values x_1 and x_2 as $x_1 = a \cdot 0 + b$ and $x_2 = a \cdot 1 + b$, for $a = x_2 - x_1$ and $b = x_1$. Here, $x = a \cdot r + b$, where

$$r = \frac{x - b}{a} = \frac{x - x_1}{x_2 - x_1}.$$

Thus, the x-scale invariance implies that

$$J(x_1, x_2, x) = J\left(0, 1, \frac{x - x_1}{x_2 - x_1}\right).$$

So, if we denote $w(r) \overset{\text{def}}{=} J(0, 1, r)$, we then conclude that

$$J(x_1, x_2, x) = w\left(\frac{x - x_1}{x_2 - x_1}\right),$$

and thus, the above expression (4.5.3) for $I(x_1, y_1, x_2, y_2, x)$ in terms of $J(x_1, x_2, x)$ takes the following simplified form:

$$I(x_1, y_1, x_2, y_2, x) = w\left(\frac{x - x_1}{x_2 - x_1}\right) \cdot y_2 + \left(1 - w\left(\frac{x - x_1}{x_2 - x_1}\right) \cdot y_2\right) \cdot y_1. \qquad (4.5.4)$$

To complete our proof, we need to show that $w(r) = r$ for all $r \in (0, 1)$.

$4°$. Let us now use consistency.

Let us take $x_1 = y_1 = 0$ and $x_2 = y_2 = 1$, then

$$I(0, 0, 1, 1, x) = w(x) \cdot 1 + (1 - w(x)) \cdot 0 = w(x).$$

Let us denote $\alpha \overset{\text{def}}{=} w(0.5)$.

By consistency, for $x = 0.25 = \dfrac{0 + 0.5}{2}$, the value $w(0.25)$ can be obtained if we apply the same interpolation procedure to $w(0) = 0$ and to $w(0.5) = \alpha$. Thus, we get

$$w(0.25) = \alpha \cdot w(0.5) + (1 - \alpha) \cdot w(0) = \alpha^2.$$

Similarly, for $x = 0.75 = \dfrac{0.5 + 1}{2}$, the value $w(0.75)$ can be obtained if we apply the same interpolation procedure to $w(0.5) = \alpha$ and to $w(1) = 1$. Thus, we get

$$w(0.75) = \alpha \cdot w(1) + (1 - \alpha) \cdot w(0.5) = \alpha \cdot 1 + (1 - \alpha) \cdot \alpha = 2\alpha - \alpha^2.$$

Finally, for $x = 0.5 = \dfrac{0.25 + 75}{2}$, the value $w(0.5)$ can be obtained if we apply the same interpolation procedure to $w(0.25) = \alpha^2$ and to $w(0.75) = 2\alpha - \alpha^2$. Thus, we get

$$w(0.5) = \alpha \cdot w(0.75) + (1 - \alpha) \cdot w(0.25) =$$

$$\alpha \cdot (2\alpha - \alpha^2) + (1 - \alpha) \cdot \alpha^2 = 3\alpha^2 - 2\alpha^3.$$

By consistency, this estimate should be equal to our original estimate $w(0.5) = \alpha$, i.e., we must have

$$3\alpha^2 - 2\alpha^3 = \alpha. \tag{4.5.5}$$

$5°$. One possible solution is to have $\alpha = 0$. In this case, we have $w(0.5) = 0$. Then, we have

$$w(0.75) = \alpha \cdot w(1) + (1 - \alpha) \cdot w(0.5) = 0,$$

and by induction, we can show that in this case, $w(1 - 2^{-n}) = 0$ for each n. In this case, $1 - 2^{-n} \to 1$, but $w(1 - 2^{-n}) \to 0$, which contradicts to the continuity requirement, according to which $w(1 - 2^{-n}) \to w(1) = 1$.

Thus, the value $\alpha = 0$ is impossible, so $\alpha \neq 0$, and we can divide both sides of the above equality (4.5.5) by α.

As a result, we get a quadratic equation

$$3\alpha - 2\alpha^2 = 1,$$

which has two solutions: $\alpha = 1$ and $\alpha = 0.5$.

$6°$. When $\alpha = 1$, we have $w(0.5) = 1$. Then, we have

$$w(0.25) = \alpha \cdot w(0.5) + (1 - \alpha) \cdot w(0) = 1,$$

and by induction, we can show that in this case, $w(2^{-n}) = 1$ for each n. In this case, $2^{-n} \to 0$, but $w(2^{-n}) \to 1$, which contradicts to the continuity requirement, according to which $w(2^{-n}) \to w(0) = 0$.

Thus, the value $\alpha = 1$ is impossible, so $\alpha = 0.5$.

7°. For $\alpha = 0.5$, we have $w(0) = 0$, $w(0.5) = 0.5$, and $w(1) = 1$. Let us prove, by induction over q, that for every binary-rational number $r = \dfrac{p}{2^q} \in [0, 1]$, we have

$$w(r) = r.$$

Indeed, the base case $q = 1$ is proven. Let us assume that we have proven it for $q - 1$, let us prove it for q. If p is even, i.e., if $p = 2k$, then $\dfrac{2k}{2^q} = \dfrac{k}{2^{q-1}}$, so the desired equality comes from the induction assumption. If $p = 2k + 1$, then

$$r = \frac{p}{2^q} = \frac{2k+1}{2^q} = 0.5 \cdot \frac{2k}{2^q} + 0.5 \cdot \frac{2 \cdot (k+1)}{2^q} = 0.5 \cdot \frac{k}{2^{q-1}} + 0.5 \cdot \frac{k+1}{2^{q-1}}.$$

By consistency, we thus have

$$w(r) = 0.5 \cdot w\left(\frac{k}{2^{q-1}}\right) + 0.5 \cdot w\left(\frac{k+1}{2^{q-1}}\right).$$

By induction assumption, we have

$$w\left(\frac{k}{2^{q-1}}\right) = \frac{k}{2^{q-1}} \text{ and } w\left(\frac{k+1}{2^{q-1}}\right) = \frac{k+1}{2^{q-1}}.$$

So, the above formula takes the form

$$w(r) = 0.5 \cdot \frac{k}{2^{q-1}} + 0.5 \cdot \frac{k+1}{2^{q-1}},$$

hence $w(r) = \dfrac{2k+1}{2^q} = r$.

The statement is proven.

8°. The equality $w(r) = r$ is true for all binary-rational numbers. Any real number x from the interval $[0, 1]$ is a limit of such numbers – namely, truncates of its infinite binary expansion. Thus, by continuity, we have $w(x) = x$ for all x.

Substituting $w(x) = x$ into the above formula (4.5.4) for $I(x_1, y_1, x_2, y_2, x)$ leads exactly to linear interpolation.

The proposition is proven.

4.6   General Case: Non-linear Effects

To measure stiffness of the compacted pavement, practitioners use the Compaction
Meter Value (CMV); a ratio between the amplitude for the first harmonic of the compactor's acceleration and the amplitude corresponding to the vibration frequency.
Numerous experiments show that CMV is highly correlated with the pavement stiffness, but as of now, there is no convincing theoretical explanation for this correlation. In this section, we provide a possible theoretical explanation for the empirical
correlation. This explanation also explains why, the stiffer the material, the more
higher-order harmonics we observe.

Results from this section first appeared in [70].

4.6.1   Compaction Meter Value (CMV) – An Empirical
Measure of Pavement Stiffness

Need to measure pavement stiffness. Road pavement must be stiff: the pavement
must remain largely unchanged when heavy vehicles pass over it.

To increase the pavement's stiffness, pavement layers are usually compacted by
the rolling compactors. In the cities, only non-vibrating compactors are used, to
avoid human discomfort caused by vibration. However, in roads outside the city
limits, vibrating compactors are used, to make compaction more efficient. In this
section, we denote the vibration frequency by f.

Compaction is applied both to the soil and to the stiffer additional pavement
material that is usually placed on top of the original soil. To check whether we need
another round of compaction and/or another layer of additional material on top, we
need to measure the current pavement stiffness.

Ideally, we should measure stiffness as we compact. In principle, we can measure
stiffness *after* each compaction cycle, but it would be definitely more efficient to
measure it *during* the compaction – this way we save time and we save additional
efforts needed for post-compaction measurements.

What we *can* rather easily measure during compaction is acceleration; it is therefore desirable to estimate the pavement stiffness based on acceleration measurements.

Compaction Meter Value (CMV). It turns out that reasonably good estimates for
stiffness can be obtained if we apply Fourier transform to the signal describing the
dependence of acceleration on time, and then evaluate *Compaction Meter Value*
(CMV), a ratio A_2/A_1 between the amplitudes corresponding to the frequencies $2f$
and f. This measure was first introduced in the late 1970s [21, 102, 103].

Numerous experiments have confirmed that CMV is highly correlated with more
direct characteristics of stiffness such as different versions of elasticity modulus; see,
e.g., [20, 55, 56, 106, 107].

CMV remains one of the main ways of estimating stiffness; see, e.g., [54].

Can we use other Fourier components? Since the use of the double-frequency component turned out to be so successful, a natural idea is to try to use other Fourier components.

It turns out that when the soil is soft (not yet stiff enough), then even the double-frequency Fourier component is not visible above noise. As the pavement becomes stiffer, we can clearly see first the first harmonic, then also higher harmonics, i.e., harmonics corresponding to $3f$, $4f$, etc.

Remaining problem. While the relation between CMV and stiffness is an empirical fact, from the theoretical viewpoint it remains somewhat a mystery: to the best of our knowledge, there is no theoretical explanation for this empirical dependence.

In this section, we attempt to provide such a theoretical explanation.

4.6.2 A Possible Theoretical Explanation of an Empirical Correlation Between CMV and Stiffness

Analysis of the problem: towards the corresponding equations. Let us start our analysis with the extreme situation when there is no stiffness at all. Crudely speaking, the complete absence of stiffness means that particles forming the soil are completely independent from each other: we can move some of them without affecting others.

In this extreme case, the displacement x_i of each particle i is determined by the Newton's equations

$$\frac{d^2 x_i}{dt^2} = \frac{1}{m_i} \cdot F_i, \tag{4.6.1}$$

where m_i is the mass of the ith particle and F_i is the force acting on this particle. For a vibrating compactor, the force F_i is sinusoidal with frequency f. Thus, the corresponding accelerations are also sinusoidal with this same frequency. In this extreme case, after the Fourier transform, we will get only one component – corresponding to the vibration frequency f.

Stiffness k means that, in addition to the external force F_i, the acceleration of each particle i is also influence by the locations of other particles x_j. For example, if we move one of the particles forming the soil, other particle move as well so that the distances between the particles remain largely the same. Thus, instead of the simple Newton's equations (4.6.1), we have more complicated equations

$$\frac{d^2 x_i}{dt^2} = \frac{1}{m_i} \cdot F_i + f_i(k, x_1, \ldots, x_N), \tag{4.6.2}$$

for some expression $f_i(k, x_1, \ldots, x_N)$.

Displacements are usually small. We consider the case when stiffness is also reasonably small. It is therefore reasonable to expand this expression in Taylor series and keep only the first few terms in this expansion.

With respect to k, in the first approximation, we just keep linear terms. With respect to x_j, it is known that the corresponding processes are observably non-linear (see, e.g., [5, 49, 64]) so we need to also take non-linear terms into account; the simplest non-linear terms are the quadratic ones, so we end up with the following approximate model:

$$\frac{d^2 x_i}{dt^2} = \frac{1}{m_i} \cdot F_i + k \cdot \sum_{j=1}^{N} a_{ij} \cdot x_j + k \cdot \sum_{j=1}^{N} \sum_{\ell=1}^{N} a_{ij\ell} \cdot x_j \cdot x_k. \qquad (4.6.3)$$

Solving the resulting equations. In general, the solution to the Eq. (4.6.3) depends on the value k: $x_i(t) = x_i(k, t)$.

When deriving the Eq. (4.6.3), we ignored terms which are quadratic (or of higher order) in terms of k. It is therefore reasonable, when looking for solutions to this equation, to also ignore terms which are quadratic (or of higher order) in k, i.e., to take

$$x_i(k, t) = x_i^{(0)}(t) + k \cdot x_i^{(1)}(t). \qquad (4.6.4)$$

If we plug in the formula (4.6.4) into the Eq. (4.6.3) and ignore terms which are quadratic in k, then we end up with the equation

$$\frac{d^2 x_i^{(0)}}{dt^2} + k \cdot \frac{d^2 x_i^{(1)}}{dt^2} = \frac{1}{m_i} \cdot F_i + k \cdot \sum_{j=1}^{N} a_{ij} \cdot x_j^{(0)} + k \cdot \sum_{j=1}^{N} \sum_{\ell=1}^{N} a_{ij\ell} \cdot x_j^{(0)} \cdot x_\ell^{(0)}. \qquad (4.6.5)$$

This formula should hold for all k, so:

- terms independent on k should be equal on both sides, and
- terms linear in k should be equal on both sides.

By equating terms in (4.6.5) that do not depend on k, we get the linear equation

$$\frac{d^2 x_i^{(0)}}{dt^2} = \frac{1}{m_i} \cdot F_i, \qquad (4.6.6)$$

which, for the sinusoidal force $F_i(t) = A_i \cdot \cos(\omega \cdot t + \Phi_i)$, has a similar sinusoidal form

$$x_i^{(0)}(t) = a_i \cdot \cos(\omega \cdot t + \varphi_i) \qquad (4.6.7)$$

for appropriate values a_i and φ_i.

By equating terms linear in k on both sides of the Eq. (4.6.5), we conclude that

$$\frac{d^2 x_i^{(1)}}{dt^2} = \sum_{j=1}^{N} a_{ij} \cdot x_j^{(0)} + \sum_{j=1}^{N} \sum_{\ell=1}^{N} a_{ij\ell} \cdot x_j^{(0)} \cdot x_\ell^{(0)}. \qquad (4.6.8)$$

For the sinusoidal expression (4.6.7) for $x_i^{(0)}$:

- linear terms $\sum_{j=1}^{N} a_{ij} \cdot x_j^{(0)}$ in the right-hand side are sinusoidal with the same angular frequency ω (i.e., with frequency f), while

- quadratic terms $\sum_{j=1}^{N} \sum_{\ell=1}^{N} a_{ij\ell} \cdot x_j^{(0)} \cdot x_\ell^{(0)}$ are sinusoids with the double angular frequency 2ω (i.e., with double frequency $2f$).

Thus, the right-hand side of the Eq. (4.6.8) is the sum of two sinusoids corresponding to frequencies f and $2f$, and so,

$$\frac{d^2 x_i}{dt^2} = \frac{d^2 x_i^{(0)}}{dt^2} + k \cdot \frac{d^2 x_i^{(1)}}{dt^2} = A_i \cdot \cos\left(\omega \cdot t + \Phi_i\right) +$$

$$k \cdot \left(A_i^{(1)} \cdot \cos\left(\omega \cdot t + \Phi_i^{(1)}\right) + A_i^{(2)} \cdot \cos\left(2\omega \cdot t + \Phi_i^{(2)}\right) \right). \tag{4.6.9}$$

The measured acceleration $a(t)$ is the acceleration of one of the points $a(t) = \dfrac{d^2 x_{i_0}(t)}{dt^2}$, thus the measured acceleration has the form

$$a(t) = A_{i_0}^{(0)} \cdot \cos\left(\omega \cdot t + \Phi_{i_0}^{(0)}\right) +$$

$$k \cdot \left(A_{i_0}^{(1)} \cdot \cos\left(\omega \cdot t + \Phi_{i_0}^{(1)}\right) + A_{i_0}^{(2)} \cdot \cos\left(2\omega \cdot t + \Phi_{i_0}^{(2)}\right) \right). \tag{4.6.10}$$

In this expression, we only have terms sinusoidal with frequency f and terms sinusoidal with frequency $2f$. Thus, in this approximation, the Fourier transform of the acceleration consists of only two components:

- a component corresponding to the main frequency f (and the corresponding angular frequency ω), and
- a component corresponding to the first harmonic $2f$, with the angular frequency 2ω.

The amplitude A_2 of the first harmonic 2ω is equal to $A_2 = k \cdot A_{i_0}^{(2)}$. The amplitude A_1 of the main frequency ω is equal to $A_1 = A_{i_0}^{(1)} + k \cdot c$ for some constant c depending on the relation between the phases. Thus, the ratio of these two amplitudes has the form

$$\frac{A_2}{A_1} = \frac{k \cdot A_{i_0}^{(2)}}{A_{i_0}^{(1)} + k \cdot c}. \tag{4.6.11}$$

In all the previous formulas, we ignored terms which are quadratic (or of higher order) in terms of k. If we perform a similar simplification in the formula (4.6.11), we conclude that

$$\frac{A_2}{A_1} = k \cdot C, \tag{4.6.12}$$

where we denoted $C \stackrel{\text{def}}{=} \dfrac{A_{i_0}^{(2)}}{A_{i_0}^{(1)}}$. In other words, we conclude that the CMV ratio is, in the first approximation, indeed proportional to stiffness.

Main conclusion. We have explained why, for reasonable small stiffness levels, we can only see two Fourier components above the noise level: the component corresponding to the vibrating frequency f and the component corresponding to the first harmonic $2f$.

We have also explained the empirical fact that the CMV – the ratio of the amplitudes of the two harmonics – is proportional to the pavement stiffness.

Case of larger stiffness: analysis and corresponding additional conclusions. When the stiffness k is sufficiently large, we can no longer ignore terms which are quadratic or of higher order in terms of k. In general, the larger the stiffness level, the more terms we need to take into account to get an accurate description of the corresponding dynamics.

Also, when the stiffness k is small, then, due to the fact that the displacements $x_i(t)$ are also reasonably small, the products of k and the terms which are, e.g., cubic in $x_j(t)$ can be safely ignored. However, when k is not very small, we need to take these terms into account as well. Using the corresponding expansion of the Eq. (4.6.3), and taking into account more terms in the expansion of $x_i(k, t)$ in k, we end up with terms which are cubic (or higher order) in terms of the ω-sinusoids $x_i^{(0)}(t)$. These terms correspond to triple, quadruple, and higher frequencies $3f$, $4f$, etc.

This is exactly what we observe: the higher the stiffness, the more higher order harmonics we see. Thus, this additional empirical fact is also theoretically explained.

Chapter 5
How General Can We Go: What Is Computable and What Is not

5.1 Formulation of the Problem

Need to take uncertainty into account when processing data: reminder. The resulting estimates are never 100% accurate:

- measurements are never absolutely accurate,
- physical models used for predictions are usually only approximate, and
- sometimes (like in quantum physics) these models only predict the probabilities of different events.

It is desirable to take this uncertainty into account when processing data.

In some cases, we know all the related probabilities; in this case, we can potentially determine the values of all statistical characteristics of interest: mean, standard deviation, correlations, etc.

Need to combine probabilistic and interval uncertainty. In most practical situations, however, we only have partial information about the corresponding probabilities. For example, for measurement uncertainties, often, the only information that we have about this uncertainty is the upper bound Δ on its absolute value; in this case, after we get a measurement result $\widetilde{X}$, the only information that we have about the actual (unknown) value of the corresponding quantity X is that it belongs to the interval $[\widetilde{X} - \Delta, \widetilde{X} + \Delta]$. We may know intervals containing the actual (unknown) cumulative distribution function, we may know bounds on moments, etc. In such situations of partial knowledge, for each statistical characteristic of interest, we can have several possible values. In such cases, we are interested in the interval of possible values of this characteristic, i.e., in the smallest and the largest possible values of this characteristic. In some cases, there are efficient algorithms for computing these intervals, in other cases, the corresponding general problem is known to be NP-hard or even not algorithmically computable; see, e.g., [37].

Studying computability – just like studying NP-hardness – is important, since it prevents us from vain attempts to solve the problem in too much generality, and

© Springer International Publishing AG, part of Springer Nature 2018 137
A. Pownuk and V. Kreinovich, *Combining Interval, Probabilistic, and Other Types of Uncertainty in Engineering Applications*, Studies in Computational Intelligence 773, https://doi.org/10.1007/978-3-319-91026-0_5

helps us concentrate on doable cases. In view of this importance, in this chapter, we describe the most general related problems which are still algorithmically solvable.

Comment. The results from this chapter first appeared in [38].

5.2   What Is Computable: A Brief Reminder

What is computable: general idea. We are interested in processing uncertainty, i.e., in dealing with a difference between the exact models of physical reality and our approximate representation of this reality. In other words, we are interested in models of physical reality.

Why do we need mathematical models in the first place? One of our main objectives is to predict the results of different actions (or the result of not performing any action). Models enable us to predict these results without the need to actually perform these actions, thus often drastically decreasing potential costs. For example, it is theoretically possible to determine the stability limits of an airplane by applying different stresses to several copies of this airplane until each copy breaks, but, if we have an adequate computer-based model, it is cheaper and faster to simulate different stresses on this model without having to destroy actual airplane frames.

From this viewpoint, a model is computable if it has algorithms that allow us to make the corresponding predictions. Let us recall how this general idea can be applied to different mathematical objects.

What is computable: case of real numbers. In modeling, real numbers usually represent values of physical quantities. This is what real numbers were originally invented for – to describe quantities like length, weight, etc., this is still one of the main practical applications of real numbers.

The simplest thing that we can do with a physical quantity is measure its value. In line with the above general idea, we can say that a real number is computable if we can predict the results of measuring the corresponding quantity.

A measurement is practically never absolutely accurate, it only produces an approximation $\tilde{x}$ to the actual (unknown) value x; see, e.g., [82]. In modern computer-based measuring instruments, such an approximate value $\tilde{x}$ is usually a binary fraction, i.e., a rational number.

For every measuring instrument, we usually know the upper bound Δ on the absolute value of the corresponding measurement error $\Delta x \stackrel{\text{def}}{=} \tilde{x} - x$: $|\Delta x| \leq \Delta$. Indeed, without such a bound, the difference Δx could be arbitrary large, and so, we would not be able to make any conclusion about the actual value x; in other words, this would be a wild guess, not a measurement.

Once we know Δ, then, based on the measurement result $\tilde{x}$, we can conclude that the actual value x is Δ-close to $\tilde{x}$: $|x - \tilde{x}| \leq \Delta$. Thus, it is reasonable to say that a real number x is computable if for every given accuracy $\Delta > 0$, we can efficiently generate a rational number that approximates x with the given accuracy.

One can easily see that it is sufficient to be able to approximate x with the accuracy 2^{-k} corresponding to k binary digits. Thus, we arrive at the following definition of a computable real number (see, e.g., [105]):

Definition 5.2.1 A real number x is called *computable* if there is an algorithm that, given a natural number k, generates a rational number r_k for which

$$|x - r_k| \leq 2^{-k}. \tag{5.2.1}$$

Comment. It is worth mentioning that not all real numbers are computable. The proof of this fact is straightforward.

Indeed, to every computable real number, there corresponds an algorithm, and different real numbers require different algorithms. An algorithm is a finite sequence of symbols. There are countably many finite sequences of symbols, so there are no more than countably many computable real numbers. And it is known that the set of all real numbers is *not* computable: this was one of the first results of set theory. Thus, there are real numbers which are not computable.

How to store a computable number in the computer. The above definition provides a straightforward way of storing a computable real number in the actual computer: namely, once we fix the accuracy 2^{-k}, all we need to store in the corresponding rational number r_k.

What is computable: case of functions from reals to reals. In the real world, there are many dependencies between the values of different quantities. Sometimes, the corresponding dependence is *functional*, in the sense that the values $x_1, \ldots, x_n$ of some quantities x_i uniquely determine the value of some other quantity y. For example, according to the Ohm's Law $V = I \cdot R$, the voltage V is uniquely determined by the values of the current I and the resistance R.

It is reasonable to say that the corresponding function $y = f(x_1, \ldots, x_n)$ is computable if, based on the results of measuring the quantities x_i, we can predict the results of measuring y. We may not know beforehand how accurately we need to measure the quantities x_i to predict y with a given accuracy k. If the original accuracy of measuring x_i is not enough, the prediction scheme can ask for more accurate measurement results. In other words, the algorithm can ask, for each pair of natural numbers $i \leq n$ and k, for a rational number r_{ik} such that $|x_i - r_{ik}| \leq 2^{-k}$. The algorithm can ask for these values r_{ik} as many times as it needs, all we require is that at the end, we always get the desired prediction. Thus, we arrive at the following definition [105]:

Definition 5.2.2 We say that a function $y = f(x_1, \ldots, x_n)$ from real numbers to real numbers is *computable* if there is an algorithm that, for all possible values x_i, given a natural number ℓ, computes a rational number s_ℓ for which

$$|f(x_1, \ldots, x_n) - s_\ell| \leq 2^{-\ell}.$$

This algorithm,

- in addition to the usual computational steps,
- can also generate *requests*, i.e., pairs of natural numbers (i, k) with $i \leq n$.

As a reply to a request, the algorithm then gets a rational number r_{ik} for which $|x_i - r_{ik}| \leq 2^{-k}$; this number can be used in further computations.

It is known that most usual mathematical functions are computable in this sense.

How to store a computable function in a computer. In contrast to the case of a computable real number, here, even if we know the accuracy $2^{-\ell}$ with which we need to compute the results, it is not immediately clear how we can store the corresponding function without explicitly storing the while algorithm.

To make storage easier, it is possible to take into account that in practice, for each physical quantity X_i, there are natural bounds $\underline{X}_i$ and $\overline{X}_i$: velocities are bounded by the speed of light, distances on Earth are bounded by the Earth's size, etc. Thus, for all practical purposes, it is sufficient to only consider values $x_i \in [\underline{X}_i, \overline{X}_i]$. It turns out that for such functions, the definition of a computable function can be simplified:

Proposition 5.2.1 *For every computable function $f(x_1, \ldots, x_n)$ on a rational-valued box $[\underline{X}_1, \overline{X}_1] \times \cdots \times [\underline{X}_n, \overline{X}_n]$, there exists an algorithm that, given a natural number ℓ, computes a natural number k such that if $|x_i - x_i'| \leq 2^{-k}$ for all i, then*

$$|f(x_1, \ldots, x_n) - f(x_1', \ldots, x_n')| \leq 2^{-\ell}. \tag{5.2.2}$$

This "ℓ to k" algorithm can be effectively constructed based on the original one.

Comment. For reader's convenience, all the proofs are placed in the special Proofs section.

Because of this result, for each ℓ, to be able to compute all the values $f(x_1, \ldots, x_n)$ with the accuracy $2^{-\ell}$, it is no longer necessary to describe the whole algorithm, it is sufficient to store finitely many rational numbers. Namely:

- We use Proposition 5.2.1 to find select a value k corresponding to the accuracy $2^{-(\ell+1)}$.
- Then, for each i, we consider a finite list of rational values

$$r_i = \underline{X}_i, \ r_i = \underline{X}_i + 2^{-k}, \ r_i = \underline{X}_i + 2 \cdot 2^{-k}, \ldots, r_i = \overline{X}_i. \tag{5.2.3}$$

- For each combination of such rational values, we use the original function's algorithm to compute the value $f(r_1, \ldots, r_n)$ with accuracy $2^{-(\ell+1)}$.

These are the values we store.

Based on these stored values, we can compute all the values of the function $f(x_1, \ldots, x_n)$ with the given accuracy $2^{-\ell}$. Specifically, for each combination of computable values $(x_1, \ldots, x_n)$, we can:

- compute 2^{-k}-close rational value $r_1, \ldots, r_n$, and then
- find, in the stored list, the corresponding approximation $\tilde{y}$ to $f(r_1, \ldots, r_n)$, i.e., the value $\tilde{y}$ for which $|f(r_1, \ldots, r_n) - \tilde{y}| \leq 2^{-(\ell+1)}$.

Let us show that this value $\widetilde{y}$ is indeed the $2^{-\ell}$-approximation to $f(x_1, \ldots, x_n)$.

Indeed, because of our choice of ℓ, from the fact that $|x_i - r_i| \le 2^{-k}$, we conclude that $|f(x_1, \ldots, x_n) - f(r_1, \ldots, r_n)| \le 2^{-(\ell+1)}$. Thus,

$$|f(x_1, \ldots, x_n) - y| \le |f(x_1, \ldots, x_n) - f(r_1, \ldots, r_n)| + |f(r_1, \ldots, r_n) - \widetilde{y}| \le$$

$$2^{-(\ell+1)} + 2^{-(\ell+1)} = 2^{-\ell}, \tag{5.2.4}$$

i.e., that the value $\widetilde{y}$ is indeed the desired $2^{-\ell}$-approximation to $f(x_1, \ldots, x_n)$.

A useful equivalent definition of a computable function. Proposition 5.2.1 allows us to use the following equivalent definition of a computable function.

Definition 5.2.2$'$ We say that a function $y = f(x_1, \ldots, x_n)$ defined on a rational-valued box $[\underline{X}_1, \overline{X}_1] \times \cdots \times [\underline{X}_n, \overline{X}_n]$ is *computable* if there exist two algorithms:

- the first algorithm, given a natural number ℓ and rational values $r_1, \ldots, r_n$, computes a $2^{-\ell}$-approximation to $f(r_1, \ldots, r_n)$;
- the second algorithm, given a natural number ℓ, computes a natural number k such that if $|x_i - x_i'| \le 2^{-k}$ for all i, then

$$|f(x_1, \ldots, x_n) - f(x_1', \ldots, x_n')| \le 2^{-\ell}. \tag{5.2.5}$$

Comment. As a corollary of Definition 5.2.2$'$, we conclude that every computable function is continuous. It should be mentioned, however, that not all continuous function is computable. For example, if a is a non-computable real number, then a linear function $f(x) = a \cdot x$ is clearly continuous but not computable. Indeed, if the function $f(x)$ was computable, we would be able to compute its value $f(1) = a$, and we know that the number a is not computable.

Not all usual mathematical functions are computable. According to Definition 5.2.2$'$, every computable function is continuous. Thus, discontinuous functions are not continuous, in particular, the following function:

Definition 5.2.3 By a *step function*, we mean a function $f(x_1)$ for which:

- $f(x_1) = 0$ for $x < 0$ and
- $f(x_1) = 1$ for $x_1 \ge 0$.

Corollary *The step function* $f(x_1)$ *is not computable.*

Comment. This corollary can be proven directly, without referring to a (rather complex) proof of Proposition 5.2.1. This direct proof is also given in the Proofs section.

Consequences for representing a probability distribution: we need to go beyond computable functions. We would like to represent a general probability distribution by its cdf $F(x)$. From the purely mathematical viewpoint, this is indeed the most general representation – as opposed, e.g., to a representation that uses a probability density function, which is not defined if we have a discrete variable.

Since the cdf $F(x)$ is a function, at first glance, it may make sense to say that the cdf is computable if the corresponding function $F(x)$ is computable. For many distributions, this definition makes perfect sense: the cdfs corresponding to uniform, Gaussian, and many other distributions are indeed computable functions.

However, for the degenerate random variable which is equal to $x = 0$ with probability 1, the cdf is exactly the step-function, and we have just proven that the step-function is not computable. Thus, we need to find an alternative way to represent cdfs, beyond computable functions.

What we do in this chapter. In this chapter, we provide the corresponding general description:

- first for case when we know the exact probability distribution, and
- then for the general case, when we only have a partial information about the probability distribution.

5.3   What We Need to Compute: A Even Briefer Reminder

The ultimate goal of all data processing is to make decision. It is known that a rational decision maker maximizes the expected value of his/her utility $u(x)$; see, e.g., [19, 46, 59, 84]. Thus, we need to be able to compute the expected values of different functions $u(x)$.

There are known procedures for eliciting from the decision maker, with any given accuracy, the utility value $u(x)$ for each x [19, 46, 59, 84]. Thus, the utility function is *computable*. We therefore need to be able to compute expected values of computable functions.

Comment. Once we are able to compute the expected values $E[u(x)]$ of different computable functions, we will thus be able to compute other statistical characteristic such as variance. Indeed, variance V can be computed as $V = E[x^2] - (E[x])^2$.

5.4   Simplest Case: A Single Random Variable

Description of the case. Let us start with the simplest case of a single random variable X. We would like to understand in what sense its cdf $F(x)$ is computable.

According to our general application-based approach to computability, this means that we would like to find out what we can compute about this random variable based on the observations.

What can we compute about $F(x)$? By definition, each value $F(x)$ is the *probability* that $X \leq x$. So, in order to decide what we can compute about the value $F(x)$, let us recall what we can compute about probabilities in general.

What can we compute about probabilities: case of an easy-to-check event. Let us first consider the simplest situation, when we consider a probability of an easy-to-check event, i.e., an event for which, from each observation, we can tell whether this event occurred or not. Such events – like observing head when tossing a coin or getting a total of seven points when throwing two dice – are what probability textbooks start with.

In general, we cannot empirically find the exact probabilities p of such an event. Empirically, we can only estimate *frequencies* f, by observing samples of different size N. It is known that for large N, the difference $d = p - f$ between the (ideal) probability and the observed frequency is asymptotically normal, with mean $\mu = 0$ and standard deviation $\sigma = \sqrt{\dfrac{p \cdot (1 - p)}{N}}$. We also know that for a normal distribution, situations when $|d - \mu| < 6\sigma$ are negligibly rare (with probability $<10^{-8}$), so for all practical purposes, we can conclude that $|f - p| \leq 6\sigma$.

If we believe that the probability of 10^{-8} is too high to ignore, we can take 7σ, 8σ, or $k_0 \cdot \sigma$ for an even larger value k_0. No matter what value k_0 we choose, for any given value $\delta > 0$, for sufficiently large N, we get $k_0 \cdot \sigma \leq \delta$.

Thus, for each well-defined event and for each desired accuracy δ, we can find the frequency f for which $|f - p| \leq \delta$. This is exactly the definition of a computable real number, so we can conclude that the probability of a well-defined event should be a computable real number.

What about the probability that $X \leq x$? The desired cdf is the probability that $X \leq x$. The corresponding event $X \leq x$ is *not* easy to check, since we do not observe the actual value X, we only observe the measurement result $\widetilde{X}$ which is close to X.

In other words, after repeating the experiment N times, instead of N actual values $X_1, \ldots, X_n$, we only know approximate values $\widetilde{X}_1, \ldots, \widetilde{X}_n$ for which

$$|\widetilde{X}_i - X_i| \leq \varepsilon \tag{5.4.1}$$

for some accuracy ε. Thus, instead of the "ideal" frequency $f = \text{Freq}(X_i \leq x)$ (which is close to the desired probability $F(x) = \text{Prob}(X \leq x)$), based on the observations, we get a slightly different frequency $f = \text{Freq}(\widetilde{X}_i \leq x)$.

What can we say about $F(x)$ based on this frequency? Since $|\widetilde{X}_i - X_i| \leq \varepsilon$, the inequality $\widetilde{X}_i \leq x$ implies that $X_i \leq x + \varepsilon$. Similarly, if if $X_i \leq x - \varepsilon$, then we can conclude that $\widetilde{X}_i \leq x$. Thus, we have:

$$\text{Freq}(X_i \leq x - \varepsilon) \leq f = \text{Freq}(\widetilde{X}_i \leq x) \leq \text{Freq}(X_i \leq x + \varepsilon). \tag{5.4.2}$$

We have already discussed that for a sufficiently large sample, frequencies are δ-close to probabilities, so we conclude that

$$\text{Prob}(X \leq x - \varepsilon) - \delta \leq f \leq \text{Prob}(\widetilde{X}_i \leq x) \leq \text{Prob}(X_i \leq x + \varepsilon) + \varepsilon. \tag{5.4.3}$$

So, we arrive at the following definition.

Definition 5.4.1 We say that a cdf $F(x)$ is *computable* if there is an algorithm that, given rational values $x, \varepsilon > 0$, and $\delta > 0$, returns a rational number f for which

$$F(x - \varepsilon) - \delta \le f \le F(x + \varepsilon) + \delta. \tag{5.4.5}$$

How to describe a computable cdf in a computer. How can we describe a computable cdf in a computer? The above definition kinds of prompts us to store the algorithm computing f, but algorithms may take a long time to compute. It is desirable to avoid such time-consuming computations and store only the pre-computed values – at least the pre-computed values corresponding to the given accuracy.

We cannot do this by directly following the above definition, since this definition requires us to produce an appropriate f for all infinitely many possible rational values x. Let us show, however, that a simple and natural modification of this idea makes storing finitely many values possible.

Indeed, for two natural numbers k and ℓ, let us take $\varepsilon_0 = 2^{-k}$ and $\delta_0 = 2^{-\ell}$. On the interval $[\underline{T}, \overline{T}]$, we then select a grid $x_1 = \underline{T}$, $x_2 = \underline{T} + \varepsilon_0, \ldots$ Due to Definition 5.4.1, for every point x_i from this grid, we can then find the value f_i for which

$$F(x_i - \varepsilon_0) - \delta_0 \le f_i \le F(x_i + \varepsilon_0) + \delta_0. \tag{5.4.5}$$

Let us also set up a grid 0, δ_0, $2\delta_0$, etc., on the interval $[0, 1]$ of possible values f_i, and instead of the original values f_i, let us store the closest values $\widetilde{f_i}$ from this grid.

Thus, for each pair (k, ℓ), we store a finite number of rational numbers $\widetilde{f_i}$ each of which take finite number of possible values (clearly not exceeding $1 + 1/\delta_0 = 2^{\ell} + 1$). Thus, for each k and ℓ, we have finitely many possible approximations of this type.

Let us show that this information is indeed sufficient to reconstruct the computable cdf, i.e., that if we have such finite-sets-of-values for all k and ℓ, then, for each rational $x, \varepsilon > 0$, and $\delta > 0$, we can algorithmically compute the value f needed in the Definition 5.4.1.

Indeed, for each ε_0 and δ_0, we can find the value x_i from the corresponding grid which is ε_0-close to x. For this x_i, we have a value $\widetilde{f_i}$ which is δ_0-close to the f_i for which

$$F(x_i - \varepsilon_0) - \delta_0 \le f_i \le F(x_i + \varepsilon_0) + \delta_0. \tag{5.4.6}$$

Thus, we have

$$F(x_i - \varepsilon_0) - 2\delta_0 \le \widetilde{f_i} \le F(x_i + \varepsilon_0) + 2\delta_0. \tag{5.4.7}$$

From $|x_i - x| \le \varepsilon_0$, we conclude that $x_i + \varepsilon_0 \le x + 2\varepsilon_0$ and $x - 2\varepsilon_0 \le x_i - \varepsilon_0$ and thus, that $F(x - 2\varepsilon_0) \le F(x_i - \varepsilon_0)$ and $F(x_i + \varepsilon_0) \le F(x + 2\varepsilon_0)$. Hence,

$$F(x - 2\varepsilon_0) - 2\delta_0 \le \widetilde{f_i} \le F(x + 2\varepsilon_0) + 2\delta_0. \tag{5.4.8}$$

So, if we take ε_0 and δ_0 for which $2\varepsilon_0 \le \varepsilon$ and $2\delta_0 \le \delta$, then we get

$$F(x - \varepsilon) \le F(x - 2\varepsilon_0) - 2\delta_0 \le \widetilde{f}_i \le F(x + 2\varepsilon_0) + 2\delta_0 \le F(x + \varepsilon) + \delta,$$
$$(5.4.9)$$

i.e., we have the desired double inequality

$$F(x - \varepsilon) - \delta \le \widetilde{f}_i \le F(x + \varepsilon) + \delta, \qquad (5.4.10)$$

with $f = \widetilde{f}_i$.

Equivalent definitions. Anyone who seriously studied mathematical papers and books have probably noticed that, in addition to definitions of different notions and theorems describing properties of these notions, these papers and books often have, for many of these notions, several different but mathematically equivalent definitions. The motivation for having several definitions is easy to understand: if we have several equivalent definitions, then in each case, instead of trying to use the original definition, we can select the one which is the most convenient to use. In view of this, let us formulate several equivalent definitions of a computable cdf.

Definition 5.4.1′ We say that a cdf $F(x)$ is *computable* if there is an algorithm that, given rational values x, $\varepsilon > 0$, and $\delta > 0$, returns a rational number f which is δ-close to $F(x')$ for some x' for which $|x' - x| \le \varepsilon$.

Proposition 5.4.1 *Definitions 5.4.1 and 5.4.1′ are equivalent to each other.*

To get the second equivalent definition, we start with the pairs $(x_i, \widetilde{f}_i)$ that we decided to use to store the computable cdf. When $f_{i+1} - f_i > \delta$, we add intermediate pairs

$$(x_i, f_i + \delta), (x_i, f_i + 2\delta), \dots, (x_i, f_{i+1}). \qquad (5.4.11)$$

We can say that the resulting finite set of pairs is (ε, δ)-*close* to the graph

$$\{(x, y) : F(x - 0) \le y \le F(x)\} \qquad (5.4.12)$$

in the following sense.

Definition 5.4.2 Let $\varepsilon > 0$ and $\delta > 0$ be two rational numbers.

- We say that pairs (x, y) and (x', y') are (ε, δ)-*close* if $|x - x'| \le \varepsilon$ and $|y - y'| \le \delta$.
- We say that the sets S and S' are (ε, δ)-*close* if:

    – for every $s \in S$, there is a (ε, δ)-close point $s' \in S'$;
    – for every $s' \in S'$, there is a (ε, δ)-close point $s \in S$.

Comment. This definition is similar to the definition of ε-closeness in *Hausdorff metric*, where the two sets S and S' are ε-close if:

- for every $s \in S$, there is a ε-close point $s' \in S'$;
- for every $s' \in S'$, there is a ε-close point $s \in S$.

Definition 5.4.1″ We say that a cdf $F(x)$ is *computable* if there is an algorithm that, given rational values $\varepsilon > 0$ and $\delta > 0$, produces a finite list of pairs which is (ε, δ)-*close* to the graph $\{(x, y) : F(x - 0) \leq y \leq F(x)\}$.

Proposition 5.4.2 *Definition 5.4.1″ is equivalent to Definitions 5.4.1 and 5.4.1′.*

Comment. Proof of Proposition 5.4.2 is similar to the above argument that our computer representation is sufficient for describing a computable cdf.

What can be computed: a positive result for the 1-D case. We are interested in computing the expected value $E_{F(x)}[u(x)]$ for computable functions $u(x)$. For this problem, we have the following result:

Theorem 5.4.1 *There is an algorithm that:*

- *given a computable cdf $F(x)$,*
- *given a computable function $u(x)$, and*
- *given (rational) accuracy $\delta > 0$,*

computes $E_{F(x)}[u(x)]$ with accuracy δ.

5.5 What if We Only Have Partial Information About the Probability Distribution?

Need to consider mixtures of probability distributions. The above result deals with the case when we have a single probability distribution, and by observing larger and larger samples we can get a better and better understanding of the corresponding probabilities. This corresponds to the ideal situation when all sub-samples have the same statistical characteristics. In practice, this is rarely the case. What we often observe is, in effect, a mixture of several samples with slightly different probabilities. For example, if we observe measurement errors, we need to take into account that a minor change in manufacturing a measuring instrument can cause a slight difference in the resulting probability distribution of measurement errors.

In such situations, instead of a *single* probability distribution, we need to consider a *set* of possible probability distributions.

Another case when we need to consider a set of distributions is when we only have partial knowledge about the probabilities. In all such cases, we need to process sets of probability distributions. To come up with an idea of how to process such sets, let us first recall how sets are dealt with in computations. For that, we will start with the simplest case: sets of numbers (or tuples).

Computational approach to sets of numbers: reminder. In the previous sections, we considered computable numbers and computable tuples (and computable functions). A number (or a tuple) corresponds to the case when we have a complete information about the value of the corresponding quantity (quantities). In practice, we often only have *partial* information about the actual value. In this case, instead of *single* value, we have a *set* of possible values. How can we represent such sets in a computer?

At first glance, this problem is complex, since there are usually infinitely many possible numbers – e.g., all numbers from an interval, and it is not clear how to represent infinitely many number in a computer – which is only capable of storing finite number of bits.

However, a more detailed analysis shows that the situation is not that hopeless: infinite number of values only appears in the idealized case when we assume that all the measurements are absolutely accurate and thus, produce the exact value. In practice, as we have mentioned, measurements have uncertainty and thus, with each measuring instrument, we can only distinguish between finitely many possible outcomes.

So, for each set S of possible values, for each accuracy ε, we can represent this set by a finite list S_ε of possible ε-accurate measurement results. This finite list has the following two properties:

- each value $s_i \in S_\varepsilon$ is the result of an ε-accurate measurement and is, thus, ε-close to some value $s \in S$;
- vice versa, each possible value $s \in S$ is represented by one of the possible measurement results, i.e., for each $s \in S$, there exists an ε-close value $s_i \in S_\varepsilon$.

Comment. An attentive reader may recognize that these two conditions have already been mentioned earlier – they correspond to ε-closeness of the sets S and S_ε in terms of Hausdorff metric.

Thus, we naturally arrive at the following definition.

Definition 5.5.1 A set S is called *computable* if there is an algorithm that, given a rational number $\varepsilon > 0$, generates a finite list S_ε for which:

- each element $s \in S$ is ε-close to some element from this list, and
- each element from this list is ε-close to some element from the set S.

Comment. In mathematics, sets which can be approximated by finite sets are known as *compact sets*. Because of this, computable sets are also known as *computable compacts*; see, e.g., [6].

So how do we describe partial information about the probability distribution. We have mentioned that for each accuracy (ε, δ), all possible probability distributions can be represented by the corresponding finite lists – e.g., if we use Definition 5.4.1″, as lists which are (ε, δ)-close to the corresponding cdf $F(x)$.

It is therefore reasonable to represent a set of probability distributions – corresponding to partial knowledge about probabilities – by finite lists of such distributions.

Definition 5.5.2 A set **S** of probability distributions is called *computable* if there is an algorithm that, given rational numbers $\varepsilon > 0$ and $\delta > 0$, generates a finite list $\mathbf{S}_{\varepsilon,\delta}$ of computable cdfs for which:

- each element $s \in \mathbf{S}$ is (ε, δ)-close to some element from this list, and
- each element from this list is (ε, δ)-close to some element from the set $\mathbf{S}$.

What can be computed? For the same utility function $u(x)$, different possible probability distributions lead, in general, to different expected values. In such a situation, it is desirable to find the *range* $E_{\mathbf{S}}[u(x)] = [\underline{E}_{\mathbf{S}}[u(x)], \overline{E}_{\mathbf{S}}[u(x)]]$ of possible values of $E_{F(x)}[u(x)]$ corresponding to all possible probability distributions $F(x) \in \mathbf{S}$:

$$\underline{E}_{\mathbf{S}}[u(x)] = \min_{F(x) \in \mathbf{S}} E_{F(x)}[u(x)]; \quad \overline{E}_{\mathbf{S}}[u(x)] = \max_{F(x) \in \mathbf{S}} E_{F(x)}[u(x)].$$

It turns out that, in general, this range is also computable:

Theorem 5.5.1 *There is an algorithm that:*

- *given a computable set $\mathbf{S}$ of probability distributions,*
- *given a computable function $u(x)$, and*
- *given (rational) accuracy $\delta > 0$,*

computes the endpoints of the range $E_{\mathbf{S}}[u(x)]$ with accuracy δ.

Comment. This result follows from Theorem 5.4.1 and from the known fact that there is a general algorithm for computing maximum and minimum of a computable function on a computable compact; see, e.g., [6].

5.6   What to Do in a General (not Necessarily 1-D) Case

Need to consider a general case. What if we have a joint distribution of several variable? A random process – i.e., a distribution on the set of functions of one variable? A random field – a probability distribution on the set of functions of several variables? A random operator? A random set?

In all these cases, we have a natural notion of a distance (metric) which is computable, so we have probability distribution on a computable metric space M.

Situations when we know the exact probability distribution: main idea. In the general case, the underlying metric space M is not always ordered, so we cannot use cdf

$$F(x) = \mathrm{Prob}(X \leq x)$$

to describe the corresponding probability distribution.

However, what we observe and measure are still numbers – namely, each measurement can be described by a computable function $g : M \to \mathbb{R}$ that maps each state $m \in M$ into a real number. By performing such measurements many times, we can get the frequencies of different values of $g(x)$. Thus, we arrive at the following definition.

Definition 5.6.1 We say that a probability distribution on a computable metric space is *computable* if there exists an algorithm, that, given:

- a computable real-valued function $g(x)$ on M, and
- rational numbers y, $\varepsilon > 0$, and $\delta > 0$,

returns a rational number f which is ε-close to the probability $\text{Prob}(g(x) \le y')$ for some y' which is δ-close to y.

How can we represent this information in a computer? Since M is a computable set, for every ε, there exists an ε-*net* $x_1, \ldots, x_n$ for M, i.e., a finite list of points for which, for every $x \in M$, there exists an ε-close point x_i from this list, thus

$$X = \bigcup_i B_\varepsilon(x_i), \text{ where } B_\varepsilon(x) \overset{\text{def}}{=} \{x' : d(x, x') \le \varepsilon\}. \tag{5.6.1}$$

For each computable element x_0, by applying the algorithm from Definition 5.6.1 to a function $g(x) = d(x, x_0)$, we can compute, for each ε_0 and δ_0, a value f which is close to $\text{Prob}(B_{\varepsilon'}(x_0))$ for some ε' which is δ_0-close to ε_0.

In particular, by taking $\delta_0 = 2^{-k}$ and $\varepsilon_0 = \varepsilon + 2 \cdot 2^{-k}$, we can find a value f' which is 2^{-k}-close to $\text{Prob}(B_{\varepsilon'}(x_0))$ for some $\varepsilon' \in [\varepsilon + 2^{-k}, \varepsilon + 3 \cdot 2^{-k}]$. Similarly, by taking $\varepsilon'_0 = \varepsilon + 5 \cdot 2^{-k}$, we can find a value f'' which is 2^{-k}-close to $\text{Prob}(B_{\varepsilon''}(x_0))$ for some $\varepsilon'' \in [\varepsilon + 4 \cdot 2^{-k}, \varepsilon + 6 \cdot 2^{-k}]$.

We know that when we have $\varepsilon < \varepsilon' < \varepsilon''$ and $\varepsilon'' \to \varepsilon$, then

$$\text{Prob}(B_{\varepsilon''}(x_0) - B_{\varepsilon'}(x_0)) \to 0, \tag{5.6.2}$$

so the values f' and f'' will eventually become close. Thus, by taking $k = 1, 2, \ldots$, we will eventually compute the number f_1 which is close to $\text{Prob}(B_{\varepsilon'}(x_1))$ for all ε' from some interval $[\underline{\varepsilon}_1, \overline{\varepsilon}_1]$ which is close to ε (and for which $\underline{\varepsilon} > \varepsilon$).

We then:

- select f_2 which is close to $\text{Prob}(B_{\varepsilon'}(x_1) \cup B_{\varepsilon'}(x_2))$ for all ε' from some interval $[\underline{\varepsilon}_2, \overline{\varepsilon}_2] \subseteq [\underline{\varepsilon}_1, \overline{\varepsilon}_1]$,
- select f_3 which is close to $\text{Prob}(B_{\varepsilon'}(x_1) \cup B_{\varepsilon'}(x_2) \cup B_{\varepsilon'}(x_3))$ for all ε' from some interval $[\underline{\varepsilon}_3, \overline{\varepsilon}_3] \subseteq [\underline{\varepsilon}_2, \overline{\varepsilon}_2]$,
- etc.

At the end, we get approximations $f_i - f_{i-1}$ to probabilities of the sets

$$S_i \overset{\text{def}}{=} B_\varepsilon(x_i) - (B_\varepsilon(x_1) \cup \cdots \cup B_\varepsilon(x_{i-1})) \tag{5.6.3}$$

for all ε from the last interval $[\underline{\varepsilon}_n, \overline{\varepsilon}_n]$.

These approximations $f_i - f_{i-1}$ form the information that we store about the probability distribution – as well as the values x_i.

What can we compute? It turns out that we can compute the expected value $E[u(x)]$ of any computable function:

Theorem 5.6.1 *There is an algorithm that:*

- *given a computable probability distribution on a computable metric space,*
- *given a computable function $u(x)$, and*
- *given (rational) accuracy $\delta > 0$,*

computes the expected value $E[u(x)]$ with accuracy δ.

What if we have a set of possible probability distributions? In the case of partial information about the probabilities, we have a set **S** of possible probability distributions.

In the computer, for any given accuracies ε and δ, each computable probability distribution is represented by the values $f_1, \ldots, f_n$. A computable set of distributions can be then defined by assuming that, for every ε and δ, instead of a single tuple $(f_1, \ldots, f_n)$, we have a *computable set* of such tuples.

In this case, similar to the 1-D situation, it is desirable to find the *range $E_\mathbf{S}[u(x)] = [\underline{E}_\mathbf{S}[u(x)], \overline{E}_\mathbf{S}[u(x)]]$* of possible values of $E_P[u(x)]$ corresponding to all possible probability distributions $P \in \mathbf{S}$:

$$\underline{E}_\mathbf{S}[u(x)] = \min_{P \in \mathbf{S}} E_{F(x)}[u(x)]; \quad \overline{E}_\mathbf{S}[u(x)] = \max_{P \in \mathbf{S}} E_{F(x)}[u(x)]. \tag{5.6.4}$$

In general, this range is also computable:

Theorem 5.6.2 *There is an algorithm that:*

- *given a computable set **S** of probability distributions,*
- *given a computable function $u(x)$, and*
- *given (rational) accuracy $\delta > 0$,*

computes the endpoints of the range $E_\mathbf{S}[u(x)]$ with accuracy δ.

Comment. Similarly to Theorem 5.5.1, this result follows from Theorem 5.6.1 and from the known fact that there is a general algorithm for computing maximum and minimum of a computable function on a computable compact [6].

5.7 Proofs

Proof of Proposition 5.2.1.

$1°$. Once we can approximate a real number x with an arbitrary accuracy, we can always find, for each k, a 2^{-k}-approximation r_k of the type $\dfrac{n_k}{2^k}$ for some integer n_k.

Indeed, we can first find a rational number r_{k+1} for which $|x - r_{k+1}| \leq 2^{-(k+1)}$, and then take $r_k = \dfrac{n_k}{2^k}$ where n_k is the integer which is the closest to the rational number $2^k \cdot r_{k+1}$. Indeed, for this closest integer, we have $|2^k \cdot r_{k+1} - n_k| \leq 0.5$. By dividing both sides of this inequality by 2^k, we get

$$|r_{k+1} - r_k| = \left|r_{k+1} - \frac{n_k}{2^k}\right| \le 2^{-(k+1)},$$

and thus, indeed,

$$|x - r_k| \le |x - r_{k+1}| + |r_{k+1} - r_k| \le 2^{-(k+1)} + 2^{-(k+1)} = 2^{-k}. \qquad (5.7.1)$$

2°. Because of Part 1 of this proof, it is sufficient to consider situations in which, as a reply to all its requests (i, k), the algorithm receives the approximate value r_{ik} of the type $\frac{n_{ik}}{2^k}$.

3°. Let us prove, by contradiction, that for given ℓ, there exists a value k_{max} that bounds, from above, the indices k in the all the requests (i, k) that this algorithm makes when computing a $2^{-\ell}$-approximation to $f(x_1, \ldots, x_n)$ on all possible inputs.

If this statement is not true, this means that for every natural number x, there exist a tuple $x^{(k)} = (x_1^{(k)}, \ldots, x_n^{(k)})$ for which this algorithm requests an approximation of accuracy at least 2^{-k} to at least one of the values $x_i^{(k)}$.

Overall, we have infinitely many tuples corresponding to infinitely many natural numbers. As a reply to each request (i, k), we get a rational number of the type $r_{ik} = \frac{n_{ik}}{2^k}$. For each natural number m, let us consider the value $\frac{p_i}{2^m}$ which is the closest to r_{ik}. There are finitely many possible tuples $(p_1, \ldots, p_n)$, so at least one of these tuples occurs infinitely many times.

Let us select such a tuple t_1 corresponding to $m = 1$. Out of infinitely many cases when we get an approximation to this tuple, we can select, on the level $m = 2$, a tuple t_2 for which we infinitely many times request the values which are 2^{-2}-close to this tuple, etc. As a result, we get a sequence of tuples t_m for which

$$|t_m - t_{m+1}| \le 2^{-m} + 2^{-(m+1)}.$$

This sequence of tuples converges. Let us denote its limit by $t = (t_1, \ldots, t_n)$. For this limit, for each k, the algorithm follows the same computation as the kth tuple and thus, will request some value with accuracy $\le 2^{-k}$. Since this is true for every k, this means that this algorithm will never stop – and we assumed that our algorithm always stops. This contradiction proves that there indeed exists an upper bound k_{max}.

4°. How can we actually find this k_{max}? For that, let us try values $m = 1, 2, \ldots$ For each m, we apply the algorithm $f(r_1, \ldots, r_n)$ to all possible combinations of values of the type $r_i = \frac{p_i}{2^m}$; in the original box, for each m, there are finitely many such tuples. For each request (i, k), we return the number of the type $\frac{n_{ik}}{2^k}$ which is the closest to t_i. When we reach the value $m = k_{max}$, then, by definition of k_{max}, this would mean that our algorithm never requires approximations which are more accurate than 2^{-m}-accurate ones.

In this case, we can then be sure that we have reached the desired value k_{max}: indeed, for all possible tuples $(x_1, \ldots, x_n)$, this algorithm will never request values beyond this mth approximation – and we have shown it for all possible combinations of such approximations. The proposition is proven.

Direct proof of the Corollary to Proposition 5.2.1. The non-computability of the step function can be easily proven by contradiction. Indeed, suppose that there exists an algorithm that computes this function. Then, for $x_1 = 0$ and $\ell = 2$, this algorithm produces a rational number s_ℓ which is 2^{-2}-close to the value $f(0) = 1$ and for which, thus, $s_\ell \geq 0.75$. This algorithm should work no matter which approximate values r_{1k} it gets – as long as these values are 2^{-k}-close to x_1. For simplicity, let us consider the case when all these approximate values are 0s: $r_{1k} = 0$.

This algorithm finishes computations in finitely many steps, during which it can only ask for the values of finitely many such approximations; let us denote the corresponding accuracies by $k_1, \ldots, k_m$, and let $K = \max(k_1, \ldots, k_m)$ be the largest of these natural numbers. In this case, all the information that this algorithm uses about the actual value x is that this value satisfies all the corresponding inequalities $|x_1 - r_{1k_j}| \leq 2^{-k_j}$, i.e., $|x_1| \leq 2^{-k_j}$. Thus, for any other value x_1' that satisfies all these inequalities, this algorithm returns the exact same value $s_\ell \geq 0.75$. In particular, this will be true for the value $x_1' = -2^{-K}$. However, for this negative value x_1', we should get $f(x_1') = 0$, and thus, the desired inequality $|f(x_1') - y_\ell| \leq 2^{-2}$ is no longer satisfied. This contradiction proves that the step function is not computable.

Proof of Proposition 5.4.1. It is easy to show that Definition 5.4.1′ implies Definition 5.4.1. Indeed, if f is δ-close to $F(x')$ for some $x' \in [x - \varepsilon, x + \varepsilon]$, i.e., if $F(x') - \delta \leq f \leq F(x') + \delta$, then, due to $x - \varepsilon \leq x' \leq x + \varepsilon$, we get $F(x - \varepsilon) \leq F(x')$ and $F(x') \leq F(x + \varepsilon)$ and thus, that

$$F(x - \varepsilon) \leq F(x') - \delta \leq f \leq F(x') + \delta \leq F(x + \varepsilon) + \delta, \qquad (5.7.2)$$

i.e., the desired inequality

$$F(x - \varepsilon) \leq f \leq F(x + \varepsilon) + \delta. \qquad (5.7.3)$$

Vice versa, let us show that Definition 5.4.1 implies Definition 5.4.1′. Indeed, we know that $F(x + \varepsilon) - F(x + \varepsilon/3) \to 0$ as $\varepsilon \to 0$. Indeed, this difference is the probability of X being in the set $\{X : x + \varepsilon/3 \leq X \leq x + \varepsilon\}$, which is a subset of the set $S_\varepsilon \stackrel{\text{def}}{=} \{X : x < X \leq x + \varepsilon\}$. The sets S_ε form a nested family with an empty intersection, thus their probabilities tend to 0 and thus, the probabilities of their subsets also tend to 0.

Due to Definition 5.4.1, for each $k = 1, 2, \ldots$, we can take $\varepsilon_k = \varepsilon \cdot 2^{-k}$ and find f_k and f_k' for which

$$F(x + \varepsilon_k/3) - \delta/4 \leq f_k \leq F(x + (2/3) \cdot \varepsilon_k) + \delta/4 \qquad (5.7.4)$$

and

$$F(x + (2/3) \cdot \varepsilon_k) - \delta/4 \leq f_k' \leq F(x + \varepsilon_k) + \delta/4. \qquad (5.7.5)$$

From these inequalities, we conclude that

$$-\delta/2 \leq f'_k - f_k \leq F(x + \varepsilon_k) - F(x + \varepsilon_k/3) + \delta/2. \qquad (5.7.6)$$

Since $F(x + \varepsilon_k) - F(x + \varepsilon_k/3) \to 0$ as $k \to \infty$, for sufficiently large k, we will have $F(x + \varepsilon_k) - F(x + \varepsilon_k/3) \leq \delta/4$ and thus, $|f'_k - f_k| \leq (3/4) \cdot \delta$. By computing the values f_k and f'_k for $k = 1, 2, \ldots$, we will eventually reach an index k for which this inequality is true. Let us show that this f_k is then δ-close to $F(x')$ for $x' = x + (2/3) \cdot \varepsilon_k$ (which is ε_k-close – and thus, ε-close – to x).

Indeed, we have

$$f_k \leq F(x + (2/3) \cdot \varepsilon_k) + \delta/4 \leq F(x + (2/3) \cdot \varepsilon_k) + \delta. \qquad (5.7.7)$$

On the other hand, we have

$$F(x + (2/3) \cdot \varepsilon_k) - \delta/4 \leq f'_k \leq f_k + (3/4) \cdot \delta \qquad (5.7.8)$$

and thus,

$$F(x + (2/3) \cdot \varepsilon_k) - \delta \leq f_k \leq F(x + (2/3) \cdot \varepsilon_k) + \delta. \qquad (5.7.9)$$

The equivalence is proven.

Proof of Theorem 5.5.1. We have shown, in Proposition 5.2.1, that every computable function $u(x)$ is computably continuous, in the sense that for every $\delta_0 > 0$, we can compute $\varepsilon > 0$ for which $|x - x'| \leq \varepsilon$ implies $|u(x) - u(x')| \leq \delta_0$.

In particular, if we take ε corresponding to $\delta_0 = 1$, and take the ε-grid $x_1, \ldots, x_i, \ldots$, then we conclude that each value $u(x)$ is 1-close to one of the values $u(x_i)$ on this grid. So, if we compute the 1-approximations $\tilde{u}_i$ to the values $u(x_i)$, then each value $u(x)$ is 2-close to one of these values $\tilde{u}_i$. Thus, $\max\limits_x |u(x)| \leq U \overset{\text{def}}{=} \max\limits_i \tilde{u}_i + 2$. So, we have a computable bound $U \geq 2$ for the (absolute value) of the computable function $u(x)$.

Let us once again use computable continuity. This time, we select ε corresponding to $\delta_0 = \delta/4$, and take an x-grid $x_1, \ldots, x_i, \ldots$ with step $\varepsilon/4$. Let G be the number of points in this grid.

According to the equivalent form (Definition 5.4.1') of the definition of computable cdf, for each of these grid points x_i, we can compute the value f_i which is $(\delta/(4U \cdot G))$-close to $F(x'_i)$ for some x'_i which is $(\varepsilon/4)$-close to x_i.

The function $u(x)$ is $(\delta/4)$-close to a piece-wise constant function $u'(x)$ which is equal to $u(x_i)$ for $x \in (x'_i, x'_{i+1}]$. Thus, their expected values are also $(\delta/4)$-close:

$$|E[u(x)] - E[u'(x)]| \leq \delta/4. \qquad (5.7.10)$$

Here, $E[u'(x)] = \sum_i u(x_i) \cdot (F(x'_{i+1}) - F(x'_i))$. But $F(x'_i)$ is $(\delta/(4U \cdot G))$-close to f_i and $F(x'_{i+1})$ is $(\delta/(4U \cdot G))$-close to f_{i+1}. Thus, each difference $F(x'_{i+1}) - F(x'_i)$ is $(\delta/(2U \cdot G))$-close to the difference $f_{i+1} - f_i$.

Since $|u(x_i)| \leq U$, we conclude that each term $u(x_i) \cdot (F(x'_{i+1}) - F(x'_i))$ is $(\delta/(2G))$-close to the computable term $u(x_i) \cdot (f_{i+1} - f_i)$. Thus, the sum of G such terms – which is equal to $E[u'(x)]$ – is $(\delta/2)$-close to the computable sum

$$\sum_i u(x_i) \cdot (f_{i+1} - f_i). \tag{5.7.11}$$

Since $E[u'(x)]$ is, in its turn, $(\delta/4)$-close to desired expected value $E[u(x)]$, we thus conclude that the above computable sum

$$\sum_i u(x_i) \cdot (f_{i+1} - f_i) \tag{5.7.12}$$

is indeed a δ-approximation to the desired expected value.

The theorem is proven.

Proof of Theorem 5.6.1. The proof is similar to the proof of Theorem 5.5.1: we approximate the function $u(x)$ by a $(\delta/2)$-close function $u'(x)$ which is piece-wise constant, namely, which is equal to a constant $u_i = u(x_i)$ on each set

$$S_i = B_\varepsilon(x_i) - (B_\varepsilon(x_1) \cup \cdots \cup B_\varepsilon(x_{i-1})). \tag{5.7.13}$$

The expected value of the function $u'(x)$ is equal to $E[u'(x)] = \sum_i u_i \cdot \text{Prob}(S_i)$.

The probabilities $\text{Prob}(S_i)$ can be computed with any given accuracy, in particular, with accuracy $\delta/(2U \cdot n)$, thus enabling us to compute $E[u'(x)]$ with accuracy $\delta/2$.

Since the functions $u(x)$ and $u'(x)$ are $(\delta/2)$-close, their expected values are also $(\delta/2)$-close. So, a $(\delta/2)$-approximation to $E[u'(x)]$ is the desired δ-approximation to $E[u(x)]$.

The theorem is proven.

5.8 Conclusions

When processing data, it is important to take into account that data comes from measurements and is, therefore, imprecise. In some cases, we know the probabilities of different possible values of measurement error – in this case, we have a probabilistic uncertainty. In other cases, we only know the upper bounds on the measurement error; in this case, we have interval uncertainty.

In general, we have a partial information about the corresponding probabilities: e.g., instead of knowing the exact values of the cumulative distribution function (cdf)

$F(x) = \text{Prob}(X \leq x)$, we only know bounds on these values – i.e., in other words, an interval containing such bounds. In such situations, we have a combination of probabilistic and interval uncertainty.

The ultimate goal of data processing under uncertainty is to have efficient algorithms for processing the corresponding uncertainty, algorithms which are as general as possible. To come up with such algorithms, it is reasonable to analyze which of the related problems are algorithmically solvable in the first place: e.g., is it possible to always compute the expected value of a given computable function?

In this chapter, we show that a straightforward (naive) formulation, most corresponding problems are not algorithmically solvable: for example, no algorithm can always, given the value x, compute the corresponding value $F(x)$ of the cdf.

However, we also show that if we instead formulate these problems in practice-related terms, then these problems become algorithmically solvable. For example, if we take into account that the value x also comes from measurement and is, thus, only known with some accuracy, it no longer make sense to look for an approximation to $F(x)$; instead, it is sufficient to look for an approximation to $F(x')$ for some x' which is close to x, and it turns out that such an approximation is always computable.

Chapter 6
Decision Making Under Uncertainty

The ultimate goal of data processing is to make appropriate decisions. In this chapter, we analyze how different types of uncertainty affect our decisions. In Sects. 6.1 and 6.2, we consider decision making under interval uncertainty: the usual case of decision making is considered in Sect. 6.1, and the case of decision making in conflict situations is analyzed in Sect. 6.2. In Sect. 6.3, we consider decision making under probabilistic uncertainty and in Sect. 6.4, we consider the case of general uncertainty.

6.1 Towards Decision Making Under Interval Uncertainty

In many practical situations, we know the exact form of the objective function, and we know the optimal decision corresponding to each values of the corresponding parameters x_i. What should we do if we do not know the exact values of x_i, and instead, we only know each x_i with uncertainty – e.g., with interval uncertainty? In this case, one of the most widely used approaches is to select, for each i, one value from the corresponding interval – usually, a midpoint – and to use the exact-case optimal decision corresponding to the selected values. Does this approach lead to the optimal solution to the interval-uncertainty problem? If yes, is selecting the midpoints the best idea? In this section, we provide answers to these questions. It turns out that the selecting-a-value-from-each-interval approach can indeed lead us to the optimal solution for the interval problem – but *not* if we select midpoints.

The results described in this section first appeared in [72, 77].

© Springer International Publishing AG, part of Springer Nature 2018   157
A. Pownuk and V. Kreinovich, *Combining Interval, Probabilistic, and Other Types of Uncertainty in Engineering Applications*, Studies in Computational Intelligence 773, https://doi.org/10.1007/978-3-319-91026-0_6

6.1.1 Formulation of the Practical Problem

Often, we know the ideal-case solution. One of the main objectives of science and engineering is to provide an optimal decision in different situations. In many practical situations, we have an algorithm that provides an optimal decision based under the condition that we know the exact values of the corresponding parameters

$$x_1, \ldots, x_n.$$

In practice, we need to take uncertainty into account. In practice, we usually know x_i with some uncertainty. For example, often, for each i, we only know an interval $[\underline{x}_i, \overline{x}_i]$ that contains the actual (unknown) value x_i; see, e.g., [82].
A problem. In the presence of such interval uncertainty, how can we find the optimal solution?

One of the most widely used approaches uses the fact that under interval uncertainty, we can implement decisions corresponding to different combinations of values $x_i \in [\underline{x}_i, \overline{x}_i]$. If this indeed a way to the solution which is optimal under interval uncertainty? If yes, which values should we choose?

Often, practitioners select the midpoints, but is this selection the best choice? These are the questions that we answer in this section.

6.1.2 Formulation of the Problem in Precise Terms

Decision making: a general description. In general, we need to make a decision $u = (u_1, \ldots, u_m)$ based on the state $x = (x_1, \ldots, x_n)$ of the system. According to decision theory, a rational person selects a decision that maximizes the value of an appropriate function known as utility; see, e.g., [19, 46, 59, 84].

We will consider situations when or each state x and for each decision u, we know the value of the utility $f(x, u)$ corresponding to us choosing u. Then, when we know the exact state x of the system, the optimal decision $u^{\text{opt}}(x)$ is the decision for which this utility is the largest possible:

$$f(x, u^{\text{opt}}(x)) = \max_u f(x, u). \tag{6.1.1}$$

Decision making under interval uncertainty. In practice, we rarely know the exact state of the system, we usually know this state with some uncertainty. Often, we do not know the probabilities of different possible states x, we only know the bounds on different parameters describing the state.

The bounds mean that for each i, instead of knowing the exact values of x_i, we only know the bounds $\underline{x}_i$ and $\overline{x}_i$ on this quantity, i.e., we only know that the actual (unknown) value x_i belongs to the interval $[\underline{x}_i, \overline{x}_i]$. The question is: what decision u should we make in this case?

We also assume that the uncertainty with which we know x is relatively small, so in the corresponding Taylor series, we can only keep the first few terms in terms of this uncertainty.

Decision making under interval uncertainty: towards a precise formulation of the problem. Because of the uncertainty with which we know the state x, for each possible decision u, we do not know the exact value of the utility, we only know that this utility is equal to $f(x, u)$ for some $x_i \in [\underline{x}_i, \overline{x}_i]$. Thus, all we know is that this utility value belongs to the interval

$$\left[\min_{x_i \in [\underline{x}_i, \overline{x}_i]} f(x_1, \ldots, x_n, u), \max_{x_i \in [\underline{x}_i, \overline{x}_i]} f(x_1, \ldots, x_n, u) \right]. \tag{6.1.2}$$

According to decision theory (see, e.g., [25, 32, 46]), if for every action a, we only know the interval $[f^-(a), f^+(a)]$ of possible values of utility, then we should select the action for which the following combination takes the largest possible value:

$$\alpha \cdot f^+(a) + (1 - \alpha) \cdot f^-(a), \tag{6.1.3}$$

where the parameter $\alpha \in [0, 1]$ describes the decision maker's degree of optimism-pessimism:

- the value $\alpha = 1$ means that the decision maker is a complete optimist, only taking into account the best-case situations,
- the value $\alpha = 0$ means that the decision maker is a complete pessimist, only taking into account the worst-case situations, and
- intermediate value $\alpha \in (0, 1)$ means that the decision maker takes into account both worst-case and best-case scenarios.

Resulting formulation of the problem. In these terms our goal is:

- given the function $f(x, u)$ and the bounds $\underline{x}$ and $\overline{x}$,
- to find the value u for which the following objective function takes the largest possible value:

$$\alpha \cdot \max_{x_i \in [\underline{x}_i, \overline{x}_i]} f(x_1, \ldots, x_n, u) + (1 - \alpha) \cdot \min_{x_i \in [\underline{x}_i, \overline{x}_i]} f(x_1, \ldots, x_n, u) \to \max_u. \tag{6.1.4}$$

Comment. The simplest case when the state x is characterized by a single parameter and when a decision u is also described by a single number, was analyzed in [72].

6.1.3   Analysis of the Problem

We assumed that the uncertainty is small, and that in the corresponding Taylor expansions, we can keep only a few first terms corresponding to this uncertainty. Therefore, it is convenient to describe this uncertainty explicitly.

Let us denote the midpoint $\dfrac{\underline{x}_i + \overline{x}_i}{2}$ of the interval $[\underline{x}_i, \overline{x}_i]$ by $\widetilde{x}_i$. Then, each value x_i from this interval can be represented as $x_i = \widetilde{x}_i + \Delta x_i$, where we denoted $\Delta x_i \stackrel{\text{def}}{=} x_i - \widetilde{x}_i$. The range of possible values of Δx_i is $[\underline{x}_i - \widetilde{x}_i, \overline{x}_i - \widetilde{x}_i] = [-\Delta_i, \Delta_i]$, where we denoted $\Delta_i \stackrel{\text{def}}{=} \dfrac{\overline{x}_i - \underline{x}_i}{2}$.

The differences Δx_i are small, so we should be able to keep only the few first terms in Δx_i.

When all the values x_i are known exactly, the exact-case optimal decision is $u^{\text{opt}}(x)$. Since uncertainty is assumed to be small, the optimal decision $u = (u_1, \ldots, u_m)$ under interval uncertainty should be close to the exact-case optimal decision $\widetilde{u} = (\widetilde{u}_1, \ldots, \widetilde{u}_m) \stackrel{\text{def}}{=} u^{\text{opt}}(\widetilde{x})$ corresponding to the midpoints of all the intervals. So, the difference $\Delta u_j \stackrel{\text{def}}{=} u_j - \widetilde{u}_j$ should also be small. In terms of Δu_j, the interval-case optimal value u_j has the form $u_j = \widetilde{u}_j + \Delta u_j$. Substituting $x_i = \widetilde{x}_i + \Delta x_i$ and $u_j = \widetilde{u}_j + \Delta u_j$ into the expression $f(x, u)$ for the utility, and keeping only linear and quadratic terms in this expansion, we conclude that

$$f(x, u) = f(\widetilde{x} + \Delta x, \widetilde{u} + \Delta u) =$$

$$f(\widetilde{x}, \widetilde{u}) + \sum_{i=1}^{n} f_{x_i} \cdot \Delta x_i + \sum_{j=1}^{m} f_{u_j} \cdot \Delta u_j +$$

$$\frac{1}{2} \cdot \sum_{i=1}^{n} \sum_{i'=1}^{n} f_{x_i x_{i'}} \cdot \Delta x_i \cdot \Delta x_{i'} + \sum_{i=1}^{n} \sum_{j=1}^{m} f_{x_i u_j} \cdot \Delta x_i \cdot \Delta u_j +$$

$$\frac{1}{2} \cdot \sum_{j=1}^{m} \sum_{j'=1}^{m} f_{u_j u_{j'}} \cdot \Delta u_j \cdot \Delta u_{j'}, \qquad (6.1.5)$$

where we denoted

$$f_{x_i} \stackrel{\text{def}}{=} \frac{\partial f}{\partial x_i}(\widetilde{x}, \widetilde{u}_j), \quad f_{u_j} \stackrel{\text{def}}{=} \frac{\partial f}{\partial u_j}(\widetilde{x}, \widetilde{u}),$$

$$f_{x_i x_{i'}} \stackrel{\text{def}}{=} \frac{\partial^2 f}{\partial x_i \partial x_{i'}}(\widetilde{x}, \widetilde{u}), \quad f_{x_i u_j} \stackrel{\text{def}}{=} \frac{\partial^2 f}{\partial x_i \partial u_j}(\widetilde{x}, \widetilde{u}),$$

$$f_{u_j u_{j'}} \stackrel{\text{def}}{=} \frac{\partial^2 f}{\partial u_j \partial u_{j'}}(\widetilde{x}, \widetilde{u}).$$

To find an explicit expression for the objective function (6.1.4), we need to find the maximum and the minimum of this objective function when u is fixed and $x_i \in [\underline{x}_i, \overline{x}_i]$, i.e., when $\Delta x_i \in [-\Delta_i, \Delta_i]$. To find the maximum and the minimum of a function of an interval, it is useful to compute its derivative. For the objective function (6.1.5), we have

$$\frac{\partial f}{\partial x_i} = f_{x_i} + \sum_{i'=1}^{n} f_{x_i x_{i'}} \cdot \Delta x_{i'} + \sum_{j=1}^{m} f_{x_i u_j} \cdot \Delta u_j. \tag{6.1.6}$$

In general, the value f_{x_i} is different from 0; possible degenerate cases when $f_{x_i} = 0$ seem to be rare. On a simple example – when the state is described by a single quantity $x = x_1$ and the decision u is also described by a single quantity $u = u_1$ – let us explain why we believe that this degenerate case can be ignored.

6.1.4 Explaining Why, In General, We Have $f_{x_i} \neq 0$

Simple case. Let us assume that the state x is the difference $x = T - T_{\text{ideal}}$ between the actual temperature T and the ideal temperature T_{ideal}. In this case, $T = T_{\text{ideal}} + x$.

Let u be the amount of degree by which we cool down the room. Then, the resulting temperature in the room is $T' = T - u = T_{\text{ideal}} + x - u$. The difference $T' - T_{\text{ideal}}$ between the resulting temperature T' and the ideal temperature T_{ideal} is thus equal to $x - u$.

It is reasonable to assume that the discomfort D depends on this difference d: $D = D(d)$. The discomfort is 0 when the difference is 0, and is positive when the difference is non-zero. Thus, if we expand the dependence $D(d)$ in Taylor series and keep only quadratic terms in this expansion $D(d) = d_0 + d_1 \cdot d + d_2 \cdot d^2$, we conclude that $d_0 = 0$, that $d_1 = 0$ (since the function $D(d)$ has a minimum at $d = 0$), and thus, that $D(d) = d_2 \cdot d^2 = d_2 \cdot (x - u)^2$, for some $d_2 > 0$.

So the utility – which is minus this discomfort – is equal to

$$f(x, u) = -d_2 \cdot (x - u)^2.$$

In this case, for each state x, the exact-case optimal decision is $u^{\text{opt}}(x) = x$. Thus, at the point where $x = x_0$ and $u = u_0 = u^{\text{opt}}(x) = x_0$, we have

$$f_x = \frac{\partial f}{\partial x} = -2d_2 \cdot (x_0 - u_0) = 0.$$

So, we have exactly the degenerate case that we were trying to avoid.

Let us make the description of this case slightly more realistic. Let us show that if we make the description more realistic, the derivative f_x is no longer equal to 0.

Indeed, in the above simplified description, we only took into account the discomfort of the user when the temperature in the room is different from the ideal. To be realistic, we need to also take into account that there is a cost $C(u)$ associated with cooling (or heating).

This cost is 0 when $u = 0$ and is non-negative when $u \neq 0$, so in the first approximation, similarly to how we described $D(d)$, we conclude that $C(u) = k \cdot u^2$, for some $k > 0$. The need to pay this cost decreases the utility function which now takes the form

$$f(x, u) = -d_2 \cdot (x - u)^2 - k \cdot u^2.$$

For this more realistic utility function, the value $u^{\mathrm{opt}}(x)$ that maximizes the utility for a given x can be found if we differentiate the utility function with respect to u and equate the derivative to 0. Thus, we get

$$2d_2 \cdot (u - x) + 2k \cdot u = 0,$$

hence $(d_2 + k) \cdot u - d_2 \cdot x = 0$, and

$$u^{\mathrm{opt}}(x) = \frac{d_2}{d_2 + k} \cdot x.$$

For $x = x_0$ and $u = u_0 = u^{\mathrm{opt}}(x_0) = \dfrac{d_2}{d_2 + k} \cdot x_0$, we thus get

$$f_x = \frac{\partial f}{\partial x} = 2(x_0 - u_0) = 2\left(x_0 - \frac{d_2}{d_2 + k} \cdot x_0\right) = \frac{2k}{d_2 + k} \cdot x_0 \neq 0.$$

So, if we make the model more realistic, we indeed get a non-degenerate case $f_x \neq 0$ that we consider in this section.

6.1.5 Analysis of the Problem (Continued)

We consider the non-degenerate case, when $f_{x_i} \neq 0$. Since we assumed that all the differences Δx_i and Δu_j are small, a linear combination of these differences is smaller than $|f_{x_i}|$. Thus, for all values Δx_i from the corresponding intervals $\Delta x_i \in [-\Delta_i, \Delta_i]$, the sign of the derivative $\dfrac{\partial f}{\partial x_i}$ is the same as the sign $s_{x_i} \overset{\text{def}}{=} \mathrm{sign}(f_{x_i})$ of the midpoint value f_{x_i}.

Hence:

- when $f_{x_i} > 0$ and $s_{x_i} = +1$, the function $f(x, u)$ is an increasing function of x_i; its maximum is attained when x_i is attained its largest possible values $\overline{x}_i$, i.e., when $\Delta x_i = \Delta_i$, and its minimum is attained when $\Delta x_i = -\Delta_i$;
- when $f_{x_i} < 0$ and $s_{x_i} = -1$, the function $f(x, u)$ is an decreasing function of x_i; its maximum is attained when x_i is attained its smallest possible values $\underline{x}_i$, i.e., when $\Delta x_i = -\Delta_i$, and its minimum is attained when $\Delta x_i = \Delta_i$.

In both cases, the maximum of the utility function $f(x, u)$ is attained when $\Delta x_i = s_{x_i} \cdot \Delta_i$ and its minimum is attained when $\Delta x_i = -s_{x_i} \cdot \Delta_i$. Thus,

$$\max_{x_i \in [\underline{x}_i, \overline{x}_i]} f(x_1, \ldots, x_n, u) = f(\widetilde{x}_1 + s_{x_1} \cdot \Delta_1, \ldots, \widetilde{x}_n + s_{x_n} \cdot \Delta_n, \widetilde{u} + \Delta u) =$$

$$f(\widetilde{x}, \widetilde{u}) + \sum_{i=1}^{n} f_{x_i} \cdot s_{x_i} \cdot \Delta_i + \sum_{j=1}^{m} f_{u_j} \cdot \Delta u_j +$$

$$\frac{1}{2} \cdot \sum_{i=1}^{n} \sum_{i'=1}^{n} f_{x_i x_{i'}} \cdot s_{x_i} \cdot s_{x_{i'}} \cdot \Delta_i \cdot \Delta_{i'} + \sum_{i=1}^{n} \sum_{j=1}^{m} f_{x_i u_j} \cdot s_{x_i} \cdot \Delta_i \cdot \Delta u_j +$$

$$\frac{1}{2} \cdot \sum_{j=1}^{m} \sum_{j'=1}^{m} f_{u_j u_{j'}} \cdot \Delta u_j \cdot \Delta u_{j'}, \tag{6.1.7}$$

and

$$\min_{x_i \in [\underline{x}_i, \overline{x}_i]} f(x_1, \ldots, x_n, u) = f(\widetilde{x}_1 - s_{x_1} \cdot \Delta_1, \ldots, \widetilde{x}_n - s_{x_n} \cdot \Delta_n, \widetilde{u} + \Delta u) =$$

$$f(\widetilde{x}, \widetilde{u}) - \sum_{i=1}^{n} f_{x_i} \cdot s_{x_i} \cdot \Delta_i + \sum_{j=1}^{m} f_{u_j} \cdot \Delta u_j +$$

$$\frac{1}{2} \cdot \sum_{i=1}^{n} \sum_{i'=1}^{n} f_{x_i x_{i'}} \cdot s_{x_i} \cdot s_{x_{i'}} \cdot \Delta_i \cdot \Delta_{i'} - \sum_{i=1}^{n} \sum_{j=1}^{m} f_{x_i u_j} \cdot s_{x_i} \cdot \Delta_i \cdot \Delta u_j +$$

$$\frac{1}{2} \cdot \sum_{j=1}^{m} \sum_{j'=1}^{m} f_{u_j u_{j'}} \cdot \Delta u_j \cdot \Delta u_{j'}. \tag{6.1.8}$$

Therefore, our objective function (6.1.4) takes the form

$$\alpha \cdot \max_{x_i \in [\underline{x}_i, \overline{x}_i]} f(x_1, \ldots, x_n, u) + (1 - \alpha) \cdot \min_{x_i \in [\underline{x}_i, \overline{x}_i]} f(x_1, \ldots, x_n, u) =$$

$$f(\widetilde{x}, \widetilde{u}) + (2\alpha - 1) \cdot \sum_{i=1}^{n} f_{x_i} \cdot s_{x_i} \cdot \Delta_i + \sum_{j=1}^{m} f_{u_j} \cdot \Delta u_j +$$

$$\frac{1}{2} \cdot \sum_{i=1}^{n} \sum_{i'=1}^{n} f_{x_i x_{i'}} \cdot s_{x_i} \cdot s_{x_{i'}} \cdot \Delta_i \cdot \Delta_{i'} + (2\alpha - 1) \cdot \sum_{i=1}^{n} \sum_{j=1}^{m} f_{x_i u_j} \cdot s_{x_i} \cdot \Delta_i \cdot \Delta u_j +$$

$$\frac{1}{2} \cdot \sum_{j=1}^{m} \sum_{j'=1}^{m} f_{u_j u_{j'}} \cdot \Delta u_j \cdot \Delta u_{j'}. \tag{6.1.9}$$

To find the interval-case optimal value $\Delta u_j^{\max} = u_j - \tilde{u}_j$ for which the objective function (6.1.4) attains its largest possible value, we differentiate the expression (6.1.9) for the objective function (6.1.4) with respect to Δu_j and equate the derivative to 0. As a result, we get:

$$f_{u_j} + (2\alpha - 1) \cdot \sum_{i=1}^{n} f_{x_i u_j} \cdot s_{x_i} \cdot \Delta_i + \sum_{j'=1}^{m} f_{u_j u_{j'}} \cdot \Delta u_{j'}^{\max} = 0. \tag{6.1.10}$$

To simplify this expression, let us now take into account that for each $x = (x_1, \ldots, x_n)$, the function $f(x, u)$ attains its maximum at the known value $u^{\mathrm{opt}}(x)$. Differentiating expression (6.1.5) with respect to Δu_j and equating the derivative to 0, we get:

$$f_{u_j} + \sum_{i=1}^{n} f_{x_i u_j} \cdot \Delta x_i + \sum_{j'=1}^{m} f_{u_j u_{j'}} \cdot \Delta^{\mathrm{opt}} u_{j'} = 0, \tag{6.1.11}$$

where we denoted $\Delta^{\mathrm{opt}} u_j \stackrel{\mathrm{def}}{=} u_j^{\mathrm{opt}} - u_j$.

For $x = \tilde{x}$, i.e., when $\Delta x_i = 0$ for all i, this maximum is attained when $u = \tilde{u}$, i.e., when $\Delta u_j = 0$ for all j. Substituting $\Delta x_i = 0$ and $\Delta u_j = 0$ into the formula (6.1.11), we conclude that $f_{u_j} = 0$ for all j. Thus, the formula (6.1.10) takes a simplified form

$$(2\alpha - 1) \cdot \sum_{i=1}^{n} f_{x_i u_j} \cdot s_{x_i} \cdot \Delta_i + \sum_{j'=1}^{m} f_{u_j u_{j'}} \cdot \Delta u_{j'}^{\max} = 0. \tag{6.1.12}$$

In general, we can similarly expand $u_j^{\mathrm{opt}}(x)$ in Taylor series and keep only a few first terms in this expansion:

$$u_j^{\mathrm{opt}}(x_1, \ldots, x_n) = u_j^{\mathrm{opt}}(\tilde{x}_1 + \Delta x_1, \ldots, \tilde{x}_n + \Delta x_n) = \tilde{u}_j + \sum_{i=1}^{n} u_{j,x_i} \cdot \Delta x_i, \tag{6.1.13}$$

where we denoted $u_{j,x_i} \stackrel{\mathrm{def}}{=} \dfrac{\partial u_j^{\mathrm{opt}}}{\partial x_i}$. Thus, for the exact-case optimal decision,

$$\Delta u_j^{\text{opt}} = u_j^{\text{opt}}(x) - \tilde{u}_j = \sum_{i=1}^n u_{j,x_i} \cdot \Delta x_i. \qquad (6.1.14)$$

Substituting this expression for Δu_j^{opt} into the formula (6.1.11), we conclude that

$$\sum_{i=1}^n f_{x_i u_j} \cdot \Delta x_i + \sum_{i=1}^n \sum_{j'=1}^m f_{u_j u_{j'}} \cdot u_{j',x_i} \cdot \Delta x_i = 0,$$

i.e., that

$$\sum_{i=1}^n \Delta x_i \cdot \left(f_{x_i u_j} + \sum_{j'=1}^m f_{u_j u_{j'}} \cdot u_{j',x_i} \right) = 0$$

for all possible combinations of Δx_i. Thus, for each i, the coefficient at Δx_i is equal to 0, i.e.,

$$f_{x_i u_j} + \sum_{j'=1}^m f_{u_j u_{j'}} \cdot u_{j',x_i} = 0$$

for all i and j, so

$$f_{x_i u_j} = -\sum_{j'-1}^m f_{u_j u_{j'}} \cdot u_{j',x_i} = 0. \qquad (6.1.15)$$

Substituting the expression (6.1.15) into the formula (6.1.12), we conclude that

$$-(2\alpha - 1) \cdot \sum_{i=1}^n \sum_{j'=1}^m f_{u_j u_{j'}} \cdot u_{j',x_i} \cdot (s_{x_i} \cdot \Delta_i) + \sum_{j'=1}^m f_{u_j u_{j'}} \cdot \Delta_{j'}^{\text{max}} = 0,$$

i.e., that

$$\sum_{j'=1}^m f_{u_j u_{j'}} \cdot \left(\Delta u_{j'}^{\text{max}} - (2\alpha - 1) \cdot \sum_{i=1}^n u_{j',x_i} \cdot (s_{x_i} \cdot \Delta_i) \right) = 0.$$

This equality is achieved when

$$\Delta u_{j'}^{\text{max}} = (2\alpha - 1) \cdot \sum_{i=1}^n u_{j',x_i} \cdot (s_{x_i} \cdot \Delta_i) \qquad (6.1.16)$$

for all j'. So, the interval-case optimal values $u_j^{\text{max}} = \tilde{u}_j + \Delta u_j^{\text{max}}$ can be described as

$$u_j^{\max} = \tilde{u}_j + (2\alpha - 1) \cdot \sum_{i=1}^{n} u_{j,x_i} \cdot (s_{x_i} \cdot \Delta_i). \tag{6.1.17}$$

In general, as we have mentioned earlier (formula (6.1.13)), we have

$$u_j^{\text{opt}}(\tilde{x}_1 + \Delta x_1, \dots, \tilde{x}_n + \Delta x_n) = \tilde{u}_j + \sum_{i=1}^{n} u_{j,x_i} \cdot \Delta x_i. \tag{6.1.18}$$

By comparing the formulas (6.1.17) and (6.1.18), we can see that $u_j^{\max}$ is equal to $u_j^{\text{opt}}(s)$ when we take $s_j = \tilde{x}_j + (2\alpha - 1) \cdot s_{x_i} \cdot \Delta_i$, i.e., that

$$u_j^{\max} = u_j^{\text{opt}}(\tilde{x}_1 + (2\alpha - 1) \cdot s_{x_1} \cdot \Delta_1, \dots, \tilde{x}_n + (2\alpha - 1) \cdot s_{x_n} \cdot \Delta_n). \tag{6.1.19}$$

Here, s_{x_i} is the sign of the derivative f_{x_i}. We have two options:

- If $f_{x_i} > 0$, i.e., if the objective function increases with x_i, then $s_{x_i} = 1$, and the expression $s_i \overset{\text{def}}{=} \tilde{x}_i + (2\alpha - 1) \cdot s_{x_i} \cdot \Delta_i$ in the formula (6.1.19) takes the form

$$s_i = \frac{\underline{x}_i + \overline{x}_i}{2} + (2\alpha - 1) \cdot \frac{\overline{x}_i - \underline{x}_i}{2} = \alpha \cdot \overline{x}_i + (1 - \alpha) \cdot \underline{x}_i. \tag{6.1.20}$$

- If $f_{x_i} < 0$, i.e., if the objective function decreases with x_i, then $s_{x_i} = -1$, and the expression $s_i = \tilde{x}_i + (2\alpha - 1) \cdot s_{x_i} \cdot \Delta_i$ in the formula (6.1.19) takes the form

$$s_i = \frac{\underline{x}_i + \overline{x}_i}{2} - (2\alpha - 1) \cdot \frac{\overline{x}_i - \underline{x}_i}{2} = \alpha \cdot \underline{x}_i + (1 - \alpha) \cdot \overline{x}_i. \tag{6.1.21}$$

So, we arrive at the following recommendation.

6.1.6 Solution to the Problem

Formulation of the problem: reminder. We assume that we know the objective function $f(x, u)$ that characterizes our gain in a situation when the actual values of the parameters are $x = (x_1, \dots, x_n)$ and we select an alternative $u = (u_1, \dots, u_m)$.

We also assume that for every state x, we know the exact-case optimal decision $u^{\text{opt}}(x)$ for which the objective function attains its largest possible value.

In a practical situation in which we only know that each value x_i is contained in an interval $[\underline{x}_i, \overline{x}_i]$, we need to find the alternative $u^{\max} = (u_1^{\max}, \dots, u_m^{\max})$ that maximizes the Hurwicz combination of the best-case and worst-case values of the objective function.

Description of the solution. The solution to our problem is to use the exact-case optimal solution $u^{\text{opt}}(s)$ corresponding to an appropriate state $s = (s_1, \dots, s_n)$.

Here, for the variables x_i for which the objective function is an increasing function of x_i, we should select

$$s_i = \alpha \cdot \overline{x}_i + (1 - \alpha) \cdot \underline{x}_i, \tag{6.1.22}$$

where α is the optimism-pessimism parameter that characterizes the decision maker.

For the variables x_i for which the objective function is a decreasing function of x_i, we should select

$$s_i = \alpha \cdot \underline{x}_i + (1 - \alpha) \cdot \overline{x}_i. \tag{6.1.23}$$

Comment. Thus, the usual selection of the midpoint s is only interval-case optimal for decision makers for which $\alpha = 0.5$; in all other cases, this selection is *not* interval-case optimal.

Discussion. Intuitively, the above solution is in good accordance with the Hurwicz criterion:

- when the objective function increases with x_i, the best possible situation corresponds to $\overline{x}_i$, and the worst possible situation corresponds to $\underline{x}_i$; thus, the Hurwicz combination corresponds to the formula (6.1.22);
- when the objective function decreases with x, the best possible situation corresponds to $\underline{x}_i$, and the worst possible situation corresponds to $\overline{x}_i$; thus, the Hurwicz combination corresponds to the formula (6.1.23).

This intuitive understanding is, however, not a proof – Hurwicz formula combines utilities, not parameter values.

6.2 What Decision to Make In a Conflict Situation Under Interval Uncertainty: Efficient Algorithms for the Hurwicz Approach

In this section, we show how to take interval uncertainty into account when solving conflict situations. Algorithms for conflict situations under interval uncertainty are known under the assumption that each side of the conflict maximizes its worst-case expected gain. However, it is known that a more general Hurwicz approach provides a more adequate description of decision making under uncertainty. In this approach, each side maximizes the convex combination of the worst-case and the best-case expected gains. In this section, we describe how to resolve conflict situations under the general Hurwicz approach to interval uncertainty.

Results presented in this section first appeared in [39].

6.2.1 Conflict Situations Under Interval Uncertainty: Formulation of the Problem and What Is Known So Far

How conflict situations are usually described. In many practical situations – e.g., in security – we have conflict situations in which the interests of the two sides are opposite. For example, a terrorist group wants to attack one of our assets, while we want to defend them. In game theory, such situations are described by *zero-sum games*, i.e., games in which the gain of one side is the loss of another side; see, e.g., [101].

To fully describe such a situation, we need to describe:

- for each possible strategy of one side and
- for each possible strategy of the other side,

what will be the resulting gain to the first side (and, correspondingly, the loss to the other side). Let us number all the strategies of the first side, and all the strategies of the second side, and let u_{ij} be the gain of the first side (negative if this is a loss). Then, the gain of the second side is $v_{ij} = -u_{ij}$.

While zero-sum games are a useful approximation, they are not always a perfect description of the situation. For example, the main objective of the terrorists may be publicity. In this sense, a small attack in the country's capital may not cause much damage but it will bring them a lot of media attention, while a more serious attack in a remote location may be more damaging to the country, but not as media-attractive. To take this difference into account, we need, for each pair of strategies (i, j), to describe both:

- the gain u_{ij} of the first side and
- the gain v_{ij} of the second side.

In this general case, we do not necessarily have $v_{ij} = -u_{ij}$ [101].

How to describe this problem in precise terms. It is a well-known fact that in conflict situations, instead of following one of the deterministic strategies, it is beneficial to select a strategy at random, with some probability. For example, if we only have one security person available and two objects to protect, then we have two deterministic strategies:

- post this person at the first objects and
- post him/her at the second object.

If we exactly follow one of these strategies, then the adversary will be able to easily attack the other – unprotected – object. It is thus more beneficial to every time flip a coin and assign the security person to one of the objects at random. This way, for each object of attack, there will be a 50% probability that this object will be defended.

In general, each corresponding strategy of the first side can be described by the probabilities $p_1, \ldots, p_n$ of selecting each of the possible strategies, so that

$$\sum_{i=1}^{n} p_i = 1. \tag{6.2.1}$$

Similarly, the generic strategy of the second side can be described by the probabilities $q_1, \ldots, q_m$ for which

$$\sum_{j=1}^{m} q_j = 1. \tag{6.2.2}$$

If the first side selects the strategy $p = (p_1, \ldots, p_n)$ and the second side selects the strategy $q = (q_1, \ldots, q_m)$, then the expected gain of the first side is equal to

$$g_1(p, q) = \sum_{i=1}^{n} \sum_{j=1}^{m} p_i \cdot q_j \cdot u_{ij}, \tag{6.2.3}$$

while the expected gain of the second side is equal to

$$g_2(p, q) = \sum_{i=1}^{n} \sum_{j=1}^{m} p_i \cdot q_j \cdot v_{ij}. \tag{6.2.4}$$

Based on this, how can we select a strategy? It is reasonable to assume that once a strategy is selected, the other side knows the corresponding probabilities – simply by observing the past history. So, if the first side selects the strategy p, the second side should select a strategy for which, under this strategy of the first side, their gain is the largest possible, i.e., the strategy $q(p)$ for which

$$g_2(p, q(p)) = \max_q g_2(p, q). \tag{6.2.5}$$

In other words,

$$q(p) = \arg\max_q g_2(p, q). \tag{6.2.6}$$

Under this strategy of the second side, the first side gains the value $g_1(p, q(p))$. A natural idea is to select the strategy p for which this gain is the largest possible, i.e., for which

$$g_1(p, q(p)) \to \max_p, \text{ where } q(p) \stackrel{\text{def}}{=} \arg\max_q g_2(p, q). \tag{6.2.7}$$

Similarly, the second side select a strategy q for which

$$g_2(p(q), q) \to \max_q, \text{ where } p(q) \stackrel{\text{def}}{=} \arg\max_p g_1(p, q). \tag{6.2.8}$$

Towards an algorithm for solving this problem. Once the strategy p of the first side is selected, the second side selects q for which its expected gain $g_2(p, q)$ is the largest possible.

The expression $g_2(p, q)$ is linear in terms of q_j. Thus, for every q, the resulting expected gain is the convex combination

$$g_2(p, q) = \sum_{j=1}^{m} q_j \cdot q_{2j}(p) \qquad (6.2.9)$$

of the gains

$$g_{2j}(p) \stackrel{\text{def}}{=} \sum_{i=1}^{n} p_i \cdot v_{ij} \qquad (6.2.10)$$

corresponding to different deterministic strategies of the second side. Thus, the largest possible gain is attained when q is a deterministic strategy.

The jth deterministic strategy will be selected by the second side if its gain at this strategy are larger than (or equal to) gains corresponding to all other deterministic strategies, i.e., under the constraint that

$$\sum_{i=1}^{n} p_i \cdot v_{ij} \geq \sum_{i=1}^{n} p_i \cdot v_{ik} \qquad (6.2.11)$$

for all $k \neq j$.

For strategies p for which the second side selects the jth response, the gain of the first side is

$$\sum_{i=1}^{n} p_i \cdot u_{ij}. \qquad (6.2.12)$$

Among all strategies p with this "j-property", we select the one for which the expected gain of the first side is the largest possible. This can be found by optimizing a linear function under constraints which are linear inequalities – i.e., by solving a *linear programming* problem. It is known that for linear programming problems, there are efficient algorithms; see, e.g., [48].

In general, we thus have m options corresponding to m different values $j = 1, \ldots, m$. Among all these m possibility, the first side should select a strategy for which the expected gain is the largest possible. Thus, we arrive at the following algorithm.

An algorithm for solving the problem. For each j from 1 to m, we solve the following linear programming problem:

$$\sum_{i=1}^{n} p_i^{(j)} \cdot u_{ij} \to \max_{p_i^{(j)}} \tag{6.2.13}$$

under the constraints

$$\sum_{i=1}^{n} p_i^{(j)} = 1, \quad p_i^{(j)} \geq 0, \quad \sum_{i=1}^{n} p_i^{(j)} \cdot v_{ij} \geq \sum_{i=1}^{n} p_i^{(j)} \cdot v_{ik} \text{ for all } k \neq j. \tag{6.2.14}$$

Out of the resulting m solutions $p^{(j)} = \left(p_1^{(j)}, \ldots, p_n^{(j)} \right)$, $1 \leq j \leq m$, we select the one for which the corresponding value $\sum_{i=1}^{n} p_i^{(j)} \cdot u_{ij}$ is the largest.

Comment. Solution is simpler in zero-sum situations, since in this case, we only need to solve one linear programming problem; see, e.g., [101].

Need for parallelization. For simple conflict situations, when each side has a small number of strategies, the corresponding problem is easy to solve.

However, in many practical situations, especially in security-related situations, we have a large number of possible deterministic strategies of each side. This happens, e.g., if we assign air marshals to different international flights. In this case, the only way to solve the corresponding problem is to perform at least some computations in parallel.

Good news is that the above problem allows for a natural parallelization: namely, all m linear programming problems can be, in principle, solved on different processors. (Not so good news is that this exhausts the possibility of parallelization: once we get to the linear programming problems, they are P-hard, i.e., provably the hardest to parallelize; see, e.g., [98].)

Need to take uncertainty into account. The above description assumed that we know the exact consequence of each combination of strategies. This is rarely the case. In practice, we rarely know the exact gains u_{ij} and v_{ij}. At best, we know the *bounds* on these gains, i.e., we know:

- the interval $[\underline{u}_{ij}, \overline{u}_{ij}]$ that contains the actual (unknown) values u_{ij}, and
- the interval $[\underline{v}_{ij}, \overline{v}_{ij}]$ that contains the actual (unknown) values v_{ij}.

It is therefore necessary to decide what to do in such situations of interval uncertainty.

How interval uncertainty is taken into account now. In the above description of a conflict situation, we mentioned that when we select the strategy p, we maximize the worst-case situation, i.e., the smallest possible gain $g_1(p, q)$ under all possible actions of the second side. It seems reasonable to apply the same idea to the case of interval uncertainty, i.e., to maximize the smallest possible gain $g_1(p, q)$ over all possible strategies of the second side *and* over all possible values $u_{ij} \in [\underline{u}_{ij}, \overline{u}_{ij}]$.

For some practically important situations, efficient algorithms for such worst-case formulation have indeed been proposed; see, e.g., [27].

Need for a more adequate formulation of the problem. In the case of adversity, it makes sense to consider the worst-case scenario: after all the adversary wants to minimize the gain of the other side.

However, in case of interval uncertainty, using the worst-case scenario may not be the most adequate idea. The problem of decision making under uncertainty, when for each alternative a, instead of the exact value $u(a)$, we only know the interval $[\underline{u}(a), \overline{u}(a)]$ of possible values of the gain, has been thoroughly analyzed.

It is known that in such situations, the most adequate decision strategy is to select an alternative a for which the following expression attains the largest possible value:

$$u^H(a) \stackrel{\text{def}}{=} \alpha \cdot \overline{u}(a) + (1 - \alpha) \cdot \underline{u}(a), \tag{6.2.15}$$

where $\alpha \in [0, 1]$ describes the decision maker's attitude; see, e.g., [25, 32, 46]. This expression was first proposed by the Nobelist Leonid Hurwicz and is thus, known as the Hurwicz approach to decision making under interval uncertainty.

In the particular case of $\alpha = 0$, this approach leads to optimizing the worst-case value $\underline{u}(a)$, but for other values α, we have different optimization problems.

What we do in this section. In this section, we analyze how to solve conflict situations under this more adequate Hurwicz approach to decision making under uncertainty.

In this analysis, we will assume that each side knows the other's parameter α, i.e., that both sides know the values α_u and α_v that characterize their decision making under uncertainty. This can be safely assumed since we can determine these values by analyzing past decisions of each side.

6.2.2 Conflict Situation Under Hurwicz-Type Interval Uncertainty: Analysis of the Problem

Once the first side selects a strategy, what should the second side do? If the first side selects the strategy p, then, for each strategy q of the second side, the actual (unknown) gain of the second side is equal to $\sum_{i=1}^{n} \sum_{j=1}^{m} p_i \cdot q_j \cdot v_{ij}$. We do not know the exact values v_{ij}, we only know the bounds $\underline{v}_{ij} \leq v_{ij} \leq \overline{v}_{ij}$. Thus, once:

- the first side selects the strategy p and
- the second side selects the strategy q,

the gain of the second side can take any value from

$$\underline{g}_2(p, q) = \sum_{i=1}^{n} \sum_{j=1}^{m} p_i \cdot q_j \cdot \underline{v}_{ij} \tag{6.2.16}$$

to

$$\bar{g}_2(p, q) = \sum_{i=1}^{n} \sum_{j=1}^{m} p_i \cdot q_j \cdot \bar{v}_{ij}. \tag{6.2.17}$$

According to Hurwicz's approach, the second side should select a strategy q for which the Hurwicz combination

$$g_2^H(p, q) \overset{\text{def}}{=} \alpha_v \cdot \bar{g}_2(p, q) + (1 - \alpha_v) \cdot \underline{g}_2(p, q) \tag{6.2.18}$$

attains the largest possible value.

Substituting the expressions (6.2.16) and (6.2.17) into the formula (6.2.18), we conclude that

$$g_2^H(p, q) = \sum_{i=1}^{n} \sum_{j=1}^{m} p_i \cdot q_j \cdot v_{ij}^H, \tag{6.2.19}$$

where we denoted

$$v_{ij}^H \overset{\text{def}}{=} \alpha_v \cdot \bar{v}_{ij} + (1 - \alpha_v) \cdot \underline{v}_{ij}. \tag{6.2.20}$$

Thus, once the first side selects its strategy p, the second side should select a strategy $q(p)$ for which the corresponding Hurwicz combination $g_2^H(p, q)$ is the largest possible, i.e., the strategy $q(p)$ for which

$$g_2^H(p, q(p)) = \max_q g_2^H(p, q). \tag{6.2.21}$$

In other words,

$$q(p) = \arg\max_q g_2^H(p, q). \tag{6.2.22}$$

Based on this, what strategy should the first side select? Under the above strategy $q = q(p)$ of the second side, the first side gains the value

$$g_1(p, q(p)) = \sum_{i=1}^{n} \sum_{j=1}^{m} p_i \cdot q_j \cdot u_{ij}. \tag{6.2.23}$$

Since we do not know the exact values u_{ij}, we only know the bounds $\underline{u}_{ij} \leq u_{ij} \leq \bar{u}_{ij}$, we therefore do not know the exact gain of the first side. All we know is that this gain will be between

$$\underline{g}_1(p, q(p)) = \sum_{i=1}^{n} \sum_{j=1}^{m} p_i \cdot q_j \cdot \underline{u}_{ij} \tag{6.2.24}$$

and

$$\overline{g}_1(p, q(p)) = \sum_{i=1}^{n} \sum_{j=1}^{m} p_i \cdot q_j \cdot \overline{u}_{ij}. \tag{6.2.25}$$

According to Hurwicz's approach, the first side should select a strategy p for which the Hurwicz combination

$$g_1^H(p, q) \stackrel{\text{def}}{=} \alpha_u \cdot \overline{g}_1(p, q(p)) + (1 - \alpha_u) \cdot \underline{g}_1(p, q(p)) \tag{6.2.26}$$

attains the largest possible value.

Substituting the expressions (6.2.24) and (6.2.25) into the formula (6.2.26), we conclude that

$$g_1^H(p, q) = \sum_{i=1}^{n} \sum_{j=1}^{m} p_i \cdot q_j \cdot u_{ij}^H, \tag{6.2.27}$$

where we denoted

$$u_{ij}^H \stackrel{\text{def}}{=} \alpha_u \cdot \overline{u}_{ij} + (1 - \alpha_u) \cdot \underline{u}_{ij}. \tag{6.2.28}$$

What strategy should the second side select? Thus, the first side will select the strategy p for which this Hurwicz combination is the largest possible, i.e., for which

$$g_1^H(p, q(p)) \to \max_p, \quad \text{where } q(p) \stackrel{\text{def}}{=} \arg\max_q g_2^H(p, q). \tag{6.2.29}$$

Similarly, the second side select a strategy q for which

$$g_2^H(p(q), q) \to \max_q, \quad \text{where } p(q) \stackrel{\text{def}}{=} \arg\max_p g_1^H(p, q). \tag{6.2.30}$$

We thus reduce the interval-uncertainty problem to the no-uncertainty case. One can easily see that the resulting optimization problem is exactly the same as in the no-uncertainty case described in Sect. 6.2.1, with the gains u_{ij}^H and v_{ij}^H described by the formulas (6.2.28) and (6.2.20).

Thus, we can apply the algorithm described in Sect. 6.2.1 to solve the interval-uncertainty problem.

6.2.3 Algorithm for Solving Conflict Situation Under Hurwicz-Type Interval Uncertainty

What is given. For every deterministic strategy i of the first side and for every deterministic strategy j of the second side, we are given:

- the interval $[\underline{u}_{ij}, \overline{u}_{ij}]$ of the possible values of the gain of the first side, and
- the interval $[\underline{v}_{ij}, \overline{v}_{ij}]$ of the possible values of the gain of the second side.

We also know the parameters α_u and α_v characterizing decision making of each side under uncertainty.

Preliminary step: forming appropriate combinations of gain bounds. First, we compute the values

$$u_{ij}^H \overset{\text{def}}{=} \alpha_u \cdot \overline{u}_{ij} + (1 - \alpha_u) \cdot \underline{u}_{ij} \qquad (6.2.31)$$

and

$$v_{ij}^H \overset{\text{def}}{=} \alpha_v \cdot \overline{v}_{ij} + (1 - \alpha_v) \cdot \underline{v}_{ij}. \qquad (6.2.32)$$

Main step. For each j from 1 to m, we solve the following linear programming problem:

$$\sum_{i=1}^{n} p_i^{(j)} \cdot u_{ij}^H \to \max_{p_i^{(j)}} \qquad (6.2.33)$$

under the constraints

$$\sum_{i=1}^{n} p_i^{(j)} = 1, \quad p_i^{(j)} \geq 0, \quad \sum_{i=1}^{n} p_i^{(j)} \cdot v_{ij}^H > \sum_{i=1}^{n} p_i^{(j)} \cdot v_{ik}^H \text{ for all } k \neq j. \quad (6.2.34)$$

Final step. Out of the resulting m solutions $p^{(j)} = \left(p_1^{(j)}, \ldots, p_n^{(j)} \right), 1 \leq j \leq m$, we select the one for which the corresponding value

$$\sum_{i=1}^{n} p_i^{(j)} \cdot u_{ij}^H \qquad (6.2.35)$$

is the largest.

Comment. In view of the fact that in the no-uncertainty case, zero-sum games are easier to process, let us consider zero-sum games under interval uncertainty. To be more precise, let us consider situations in which possible values v_{ij} are exactly values $-u_{ij}$ for possible u_{ij}:

$$[\underline{v}_{ij}, \overline{v}_{ij}] = \{-u_{ij} : \underline{u}_{ij} \in [\underline{u}_{ij}, \overline{u}_{ij}]\}. \qquad (6.2.36)$$

One can easily see (see, e.g., [26, 58]) that this condition is equivalent to

$$\underline{v}_{ij} = -\overline{u}_{ij} \text{ and } \overline{v}_{ij} = -\underline{u}_{ij}. \qquad (6.2.37)$$

In this case, we have

$$v_{ij}^H = \alpha_v \cdot \overline{v}_{ij} + (1 - \alpha_v) \cdot \underline{v}_{ij} = \alpha_v \cdot (-\underline{u}_{ij}) + (1 - \alpha_v) \cdot (-\overline{u}_{ij}), \quad (6.2.38)$$

and thus,

$$v_{ij}^H = -((1 - \alpha_v) \cdot \overline{u}_{ij} + \alpha_v \cdot \underline{u}_{ij}). \quad (6.2.39)$$

By comparing this expression with the formula (6.2.31) for u_{ij}^H, we can conclude that the resulting game is zero-sum (i.e., $v_{ij}^H = -u_{ij}^H$) only when $\alpha_u = 1 - \alpha_v$.

In all other cases, even if we start with a zero-sum interval-uncertainty game, the no-uncertainty game to which we reduce that game will *not* be zero-sum – and thus, the general algorithm will be needed, without a simplification that is available for zero-sum games.

6.2.4 Conclusion

In this section, we show how to take interval uncertainty into account when solving conflict situations.

Algorithms for conflict situations under interval uncertainty are known under the assumption that each side of the conflict maximizes its worst-case expected gain. However, it is known that a more general Hurwicz approach provides a more adequate description of decision making under uncertainty. In this approach, each side maximizes the convex combination of the worst-case and the best-case expected gains.

In this section, we describe how to resolve conflict situations under the general Hurwicz approach to interval uncertainty.

6.3 Decision Making Under Probabilistic Uncertainty: Why Unexpectedly Positive Experiences Make Decision Makers More Optimistic

Experiments show that unexpectedly positive experiences make decision makers more optimistic. However, there seems to be no convincing explanation for this experimental fact. In this section, we show that this experimental phenomenon can be naturally explained within the traditional utility-based decision theory.

Results described in this section first appeared in [78].

6.3.1 Formulation of the Problem

Experimental phenomenon. Experiments show that unexpectedly positive experiences make decision makers more optimistic. This was first observed on an experiment with rats [88]: rats like being tickled, and tickled rats became more optimistic. Several later papers showed that the same phenomenon holds for other decision making situations as well; see, e.g., [23, 65].

Similarly, decision makers who had an unexpectedly negative experiences became more pessimistic; see, e.g., [66].

Why: a problem. There seems to be no convincing explanation for this experimental fact.

What we do in this section. In this section, we show that this experimental phenomenon can be naturally explained within the traditional utility-based decision theory.

6.3.2 Formulating the Problem in Precise Terms

In precise terms, what does it mean to becomes more optimistic or less optimistic? The traditional utility-based decision theory describes the behavior of a rational decision maker in situations in which we know the probabilities of all possible consequences of each action; see, e.g., [19, 46, 59, 84]. This theory shows that under this rationality assumption, preferences of a decision maker can be described by a special function $U(x)$ called *utility function*, so that a rational decision maker selects an alternative a that maximizes the expected value $u(a)$ of the utility.

In this case, there is no such thing as optimism or pessimism: we just select the alternative which we know is the best for us.

The original theory describes the behavior of decision makers in situations in which we know the probability of each possible consequence of each action. In practice, we often have only *partial* information about these probabilities – and sometimes, no information at all. In such situations, there are several possible probability distributions consistent with our knowledge. For different distributions, we have, in general, different values of the expected utility. As a result, for each alternative, instead of the exact value of the expected utility, we have an *interval* $[\underline{u}(a), \overline{u}(a)]$ of possible values of $u(a)$. How can we make a decision based on such intervals?

In this case, natural rationality ideas lead to the conclusion that a decision should select an alternative a for which, for some real number $\alpha \in [0, 1]$, the combination

$$u(a) = \alpha \cdot \overline{u}(a) + (1 - \alpha) \cdot \underline{u}(a)$$

is the largest possible; see, e.g., [32]. This idea was first proposed by the Nobelist Leo Hurwicz in [25].

The selection of α, however, depends on the person. The value $\alpha = 1$ means that the decision maker only takes into account the best possible consequences, and completely ignores possible less favorable situations. In other words, the values $\alpha = 1$ corresponds to complete optimism.

Similarly, the value $\alpha = 0$ means that the decision maker only takes into account the worst possible consequences, and completely ignores possible more favorable situations. In other words, the value $\alpha = 0$ corresponds to complete pessimism.

Intermediate values α mean that we take into account both positive and negative possibilities. The larger α, the close this decision maker to complete optimism. The smaller α, the closer the decision maker to complete pessimism. Because of this, the parameter α – known as the *optimism-pessimism parameter* – is a numerical measure of the decision maker's optimism.

In these terms:

- becoming more optimistic means that the value α increases, and
- becoming less optimistic means that the value α increases.

Thus, the above experimental fact takes the following precise meaning:

- if a decision maker has unexpectedly positive experiences, then this decision maker's α increases;
- if a decision maker has unexpectedly negative experiences, then this decision maker's α decreases.

This is the phenomenon that we need to explain.

6.3.3 Towards the Desired Explanation

Optimism-pessimism parameter α can be naturally interpreted as the subjective probability of positive outcome. The value α means that the decision maker selects an alternative a for which the value $\alpha \cdot \overline{u}(a) + (1 - \alpha) \cdot \underline{u}(a)$ is the largest possible.

Here, the value $\overline{u}(a)$ corresponds to the positive outcome, and the value $\underline{u}(a)$ corresponds to negative outcome.

For simplicity, let us consider the situation when we have only two possible outcomes:

- the positive outcome, with utility $\overline{u}(a)$, and
- the negative outcome, with utility $\underline{u}(a)$.

A traditional approach to decision making, as we have mentioned, assumes that we know the probabilities of different outcomes. In this case of uncertainty, we do not know the actual (objective) probabilities, but we can always come up with estimated (subjective) ones.

Let us denote the subjective probability of the positive outcome by p_+. Then, the subjective probability of the negative outcome is equal to $1 - p_+$. The expected utility is equal to $p_+ \cdot \overline{u}(a) + (1 - p_+) \cdot \underline{u}(a)$.

This is exactly what we optimize when we use Hurwicz's approach, with $\alpha = p_+$. Thus, the value α can be interpreted as the subjective probability of the positive outcome.

A new reformulation of our problem. In these terms, the above experimental phenomenon takes the following form:

- unexpectedly positive experiences increase the subjective probability of a positive outcome, while
- unexpectedly negative experiences decrease the subjective probability of a positive outcome.

To explain this phenomenon, let us recall where subjective probabilities come from.

Where subjective probabilities come from? A natural way to estimate the probability of an event is to consider all situations in which this event could potentially happen, and then take the frequency of this event – i.e., the ratio n/N of the number of times n when it happens to the overall number N of cases – as the desired estimate for the subjective probability. For example, if we flip a coin 10 times and it fell heads 6 times out of 10, we estimate the probability of the coin falling heads as 6/10.

Let us show that this leads to the desired explanation.

Resulting explanation. Suppose that a decision maker had n positive experiences in the past N situations. Then, the decision maker's subjective probability of a positive outcome is $p_+ = n/N$.

Unexpectedly positive experiences means that we have a series of new experiments, in which the fraction of positive outcomes was higher than the expected frequency p_+. In other words, unexpectedly positive experiences means that $n'/N' > p$, where N' is the overall number of new experiences, and n' is the number of those new experiences in which the outcome turned out to be positive.

How will these new experiences change the decision maker's subjective probability? Now, the decision maker has encountered overall $N + N'$ situations, of which $n + n'$ were positive. Thus, the new subjective probability p'_+ is equal to the new ratio $p'_+ = \dfrac{n + n'}{N + N'}$. Here, by definition of p_+, we have

$$n = p_+ \cdot N$$

and, due to unexpected positiveness of new experiences, we have $n' > p_+ \cdot N'$. By adding this inequality and the previous equality, we conclude that

$$n + n' > p_+ \cdot (N + N'),$$

i.e., that

$$p'_+ = \frac{n + n'}{N + N'} > p_+.$$

In other words, unexpectedly positive experiences increase the subjective probability of a positive outcome.

As we have mentioned, the subjective probability of the positive outcome is exactly the optimism-pessimism parameter α. Thus:

- the original subjective probability p_+ is equal to the original optimism-pessimism parameter α, and
- the new subjective probability p'_+ is equal to the new optimism-pessimism parameter α'.

So, the inequality $p'_+ > p_+$ means that $\alpha' > \alpha$, i.e., that unexpectedly positive experiences make the decision maker more optimistic. This is exactly what we wanted to explain.

Similarly, if we had unexpectedly negative experiences, i.e., if we had

$$n' < p_+ \cdot N',$$

then we similarly get $n + n' < p_+ \cdot (N + N')$ and thus,

$$p'_+ = \frac{n + n'}{N + N'} < p_+$$

and $\alpha' < \alpha$. So, we conclude that unexpectedly negative experiences make the decision maker less optimistic. This is also exactly what we observe. So, we have the desired explanation.

6.4   Decision Making Under General Uncertainty

There exist techniques for decision making under specific types of uncertainty, such as probabilistic, fuzzy, etc. Each of the corresponding ways of describing uncertainty has its advantages and limitations. As a result, new techniques for describing uncertainty appear all the time. Instead of trying to extend the existing decision making idea to each of these new techniques one by one, we attempt to develop a general approach that would cover all possible uncertainty techniques.

Results described in this section first appeared in [71].

6.4.1   Formulation of the Problem

Need for decision making under uncertainty. The ultimate goal of science and engineering is to make decisions, i.e., to select the most appropriate action.

Situations when we have full information about possible consequences of each action are rare. Usually, there is some uncertainty. It is therefore important to make decisions under uncertainty.

There are many different techniques for describing uncertainty. There are many different techniques for describing uncertainty: probabilistic, fuzzy (see, e.g., [28, 61, 108]), possibilistic, interval-valued or, more generally, type-2 fuzzy (see, e.g., [51, 52]), complex-valued fuzzy [13], etc. For many of these techniques, there are known methods for decision making under the corresponding uncertainty.

All the current techniques for describing uncertainty have their advantages and their limitations. Because of the known limitations, new – more adequate – techniques for describing uncertainty appear all the time. For each of these techniques, we need to understand how to make decisions under the corresponding uncertainty.

A problem that we try to solve in this section. At present, this understanding mostly comes technique-by-technique. A natural question is: can we develop a general framework that would allow us to make decision under general uncertainty?

The main objective of this section is to develop such a general formalism.

Towards a precise formulation of the problem. Let us start with a monetary problem. Suppose that we need to make a financial decision, such as investing a given amount of money in a certain financial instrument (such as shares or bonds).

If we knew the exact consequences of this action, then we would know exactly how much money we will have after a certain period of time. This happens, e.g., if we simply place the given amount in a saving account with a known interest rate.

In most situations, however, we are uncertainty of the possible financial consequences of this action. In other words, for each investment scheme, there are several possible consequences, with monetary amounts $x_1, \ldots, x_n$. By using an appropriate uncertainty technique, we can describe our degree of certainty that the ith alternative is possible by the corresponding value μ_i. Depending on the formalism for describing uncertainty,

- a value μ_i can be a number – e.g., when we use probabilistic or fuzzy uncertainty,
- it can be an interval – when we use interval-valued fuzzy,
- it can be a complex number – if we use complex-valued fuzzy,
- it can be a fuzzy set – if we use type-2 fuzzy techniques, etc.

For another investment scheme, we can have n' different possible consequences, with monetary values $x'_1, \ldots, x'_{n'}$ and degrees of certainty $\mu'_1, \ldots, \mu'_{n'}$.

To make a decision, we need to compare this investment, in particular, with situations like placing money in a saving account, in which we simply get a fixed amount of money after the same period of time.

- If this fixed amount of money is too small, then investing in an uncertain financial instrument is clearly better.
- If this fixed amount of money is sufficiently large, then getting this fixed amount of money is clearly better than investing in an uncertain financial instrument.

There should be a threshold value of the fixed amount at which we go from the instrument being preferable to a fixed amount being preferable. This threshold fixed amount of money is thus equivalent, to the user, to the investment in an uncertain instrument.

So, for each uncertain investment, in which we get:

- the amount x_1 with degree of possibility μ_1,
- the amount x_2 with degree of possibility μ_2,
- ...,
- amount x_n with degree of possibility μ_n,

we have an equivalent amount of money. We will denote this equivalent amount of money by $f(x_1, \ldots, x_n, \mu_1, \ldots, \mu_n)$.

Our goal is to find out how this equivalent amount of money depends on the values x_i and μ_i. Once we know the equivalent amount of money corresponding to each uncertain investment, we will be able to select the best of the possible investments: namely, it is natural to select the investment for which the corresponding equivalent amount of money is the largest possible.

What about non-financial decision making situations? It is known (see, e.g., [19, 46, 59, 84]) that decisions of a rational person can be described as optimizing a certain quantity called *utility*.

Thus, in general, we have the following problem: for each uncertain situation, in which we get:

- utility x_1 with degree of possibility μ_1,
- utility x_2 with degree of possibility μ_2,
- ...,
- utility x_n with degree of possibility μ_n,

we have an equivalent utility value. We will denote this equivalent utility value by $f(x_1, \ldots, x_n, \mu_1, \ldots, \mu_n)$.

Our goal is thus to find out how this equivalent utility value depends on the values x_i and μ_i. Once we know the equivalent utility value corresponding to each possible decision, we will be able to select the best of the possible decisions: namely, it is natural to select the decision for which the corresponding equivalent utility value is the largest possible.

Comment. In the following text, to make our thinking as understandable as possible, we will most talk about financial situations – since it is easier to think about money than about abstract utilities. However, our reasoning is applicable to utilities as well.

6.4.2   Analysis of the Problem

First reasonable assumption: additivity. We are interested in finding a function $f(x_1, \ldots, x_n, \mu_1, \ldots, \mu_n)$ of $2n$ variables.

Suppose that the money that we get from the investment comes in two consequent payments. In the ith alternative, we first get the amount x_i, and then – almost immediately – we also get the amount y_i.

We can consider the resulting investment in two different ways. First, we can simply ignore the fact that the money comes in two installments, and just take into account that in each alternative i, we get the amount $x_i + y_i$. This way, the equivalent amount of money is equal to

$$f(x_1 + y_1, \ldots, x_n + y_n, \mu_1, \ldots, \mu_n).$$

Alternatively, we can treat both installments separately:

- in the first installment, we get x_i with uncertainty μ_i,
- in the second installment, we get y_i with uncertainty μ_i.

Thus:

- the first installment is worth the amount $f(x_1, \ldots, x_n, \mu_1, \ldots, \mu_n)$, and
- the second installment is worth the amount $f(y_1, \ldots, y_n, \mu_1, \ldots, \mu_n)$.

The overall benefit is the sum of the amounts corresponding to both installments. So, in this way of description, the overall money value of the original investment is equal to the sum of the money values of the two installments:

$$f(x_1, \ldots, x_n, \mu_1, \ldots, \mu_n) + f(y_1, \ldots, y_n, \mu_1, \ldots, \mu_n).$$

The equivalent benefit of the investment should not depend on the way we compute it, so the two estimates should be equal:

$$f(x_1 + y_1, \ldots, x_n + y_n, \mu_1, \ldots, \mu_n) =$$

$$f(x_1, \ldots, x_n, \mu_1, \ldots, \mu_n) + f(y_1, \ldots, y_n, \mu_1, \ldots, \mu_n).$$

Functions satisfying this property are known as *additive*. Thus, we can say that for each combination of values $\mu_1, \ldots, \mu_n$, the dependence on $x_1, \ldots, x_n$ is additive.

Second reasonable assumption: bounds. No matter what happens, we get at least $\min_i x_i$ and at most $\max_i x_i$. Thus, the equivalent benefit of an investment cannot be smaller than $\min_i x_i$ and cannot be larger than $\max_i x_i$:

$$\min_i x_i \leq f(x_1, \ldots, x_n, \mu_1, \ldots, \mu_n) \leq \max_i x_i.$$

What we can conclude form the first two assumptions. It is known (see, e.g., [2]) that every bounded additive function is linear, i.e., that we have

$$f(x_1, \ldots, x_n, \mu_1, \ldots, \mu_n) = \sum_{i=1}^{n} c_i(\mu_1, \ldots, \mu_n) \cdot x_i.$$

So, instead of a function of $2n$ variables, we now have a simpler task for finding n functions $c_i(\mu_1, \ldots, \mu_n)$ of n variables.

Nothing should depend on the ordering of the alternatives. The ordering of the alternatives is arbitrary, so nothing should change if we change this ordering. For example, if we swap the first and the second alternatives, then instead of

$$c_1(\mu_1, \mu_2, \ldots) \cdot x_1 + c_2(\mu_1, \mu_2, \ldots) \cdot x_2 + \cdots$$

we should have

$$c_2(\mu_2, \mu_1, \ldots) \cdot x_1 + c_1(\mu_2, \mu_1, \ldots) \cdot x_2 + \cdots$$

These two expression must coincide, so the coefficients at x_1 must coincide, and we must have

$$c_2(\mu_2, \mu_1, \ldots) = c_1(\mu_1, \mu_2, \ldots).$$

In general, we should thus have

$$c_i(\mu_1, \ldots, \mu_n) = \mu_1(\mu_i, \mu_1, \ldots \mu_{i-1}, \mu_{i+1}, \ldots, \mu_n).$$

Thus, the above expression should have the form

$$f(x_1, \ldots, x_n) = \sum_{i=1}^{n} c_1(\mu_i, \mu_1, \ldots, \mu_{i-1}, \mu_{i+1}, \ldots, \mu_n) \cdot x_i.$$

Now, the problem is to find a single function $c_1(\mu_1, \ldots, \mu_n)$ of n variables.

Combining alternatives with the same outcomes. Based on the above formula, the value $c_1(\mu_1, \mu_2, \ldots, \mu_n)$ corresponds to $f(1, 0, \ldots, 0)$, i.e., to a situation when we have:

- the value 1 with degree of possibility μ_1,
- the value 0 with degree of possibility μ_2,
- ...,
- the value 0 with degree of possibility μ_n.

In alternatives 2 through n, we have the same outcome 0, so it makes sense to consider them as a single alternative. To find the degree of possibility of this combined alternatives, we need to apply some "or"-operation $\oplus$ to the degrees of possibility $\mu_2, \ldots, \mu_n$ of individual alternatives.

For probabilities, this combination operation is simply the sum $a \oplus b = a + b$, for fuzzy, it is a t-conorm, etc. In general, the degree of certainty of the combined alternative is equal to $\mu_2 \oplus \cdots \oplus \mu_n$. Thus, the equivalent value of this situation is equal to $c_1(\mu_1, \mu_2 \oplus \cdots \oplus \mu_n)$. So, we have

$$c_1(\mu_1, \mu_2, \ldots, \mu_n) = c_1(\mu_1, \mu_2 \oplus \cdots \oplus \mu_n),$$

and the above expression for the equivalent benefit takes the following form

$$f(x_1, \ldots, x_n) = \sum_{i=1}^{n} c_1(\mu_i, \mu_1 \oplus \mu_{i-1} \oplus \mu_{i+1} \oplus \cdots \oplus \mu_n) \cdot x_i.$$

Now, the problem is to find a single function $c_1(\mu_1, \mu_2)$ of two variables.

Let us simplify this problem even further.

Yet another reasonable requirement. Let us consider a situation in which we have three alternatives, i.e., in which, we get:

- the amount x_1 with degree of possibility μ_1,
- the amount x_2 with degree of possibility μ_2, and
- the amount x_3 with degree of possibility μ_3.

According to the above formula, for this situation, the equivalent benefit is equal to

$$c_1(\mu_1, \mu_2 \oplus \mu_3) \cdot x_1 + c_1(\mu_2, \mu_1 \oplus \mu_3) \cdot x_2 + c_1(\mu_3, \mu_1 \oplus \mu_2) \cdot x_2.$$

On the other hand, we can consider an auxiliary situation A in which we get:

- the amount x_1 with degree of possibility μ_1 and
- the amount x_2 with the degree of possibility μ_2.

This situation is equivalent to the amount

$$x_A = c_1(\mu_1, \mu_2) \cdot x_1 + c_1(\mu_2, \mu_1) \cdot x_2,$$

and the degree of possibility of this auxiliary situation can be obtained by applying the corresponding "or"-operation to the degrees μ_1 and μ_2 and is thus, equal to $\mu_A = \mu_1 \oplus \mu_2$.

By replacing the first two alternatives in the original 3-alternative situation with the equivalent alternative, we get the equivalent situation, in which we get:

- the value x_A with degree of possibility μ_A and
- the value x_3 with degree of possibility μ_3.

For this equivalent situation, the equivalent amount is equal to

$$c_1(\mu_A, \mu_3) \cdot x_A + c_1(\mu_3, \mu_A) \cdot x_3.$$

Substituting the expressions for x_A and μ_A into this formula, we conclude that the equivalent amount is equal to

$$c_1(\mu_1 \oplus \mu_2, \mu_3) \cdot (c_1(\mu_1, \mu_2) \cdot x_1 + c_1(\mu_2, \mu_2) \cdot x_2) + c_1(\mu_3, \mu_1 \oplus \mu_2) \cdot x_3 =$$

$$c_1(\mu_1 \oplus \mu_2, \mu_3) \cdot c_1(\mu_1, \mu_2) \cdot x_1 + c_1(\mu_1 \oplus \mu_2, \mu_3) \cdot c_1(\mu_2, \mu_2) \cdot x_2 +$$

$$c_1(\mu_3, \mu_1 \oplus \mu_2) \cdot x_3.$$

We get two expressions for the same equivalent amount. These expressions must coincide. This means, in particular, that the coefficients at x_1 at both expressions must coincide, i.e., that we should have

$$c_1(\mu_1, \mu_2 \oplus \mu_3) = c_1(\mu_1 \oplus \mu_2, \mu_3) \cdot c_1(\mu_1, \mu_2).$$

What can we extract from this requirement. Let us consider an auxiliary function $c(a, b) \stackrel{\text{def}}{=} c_1(a, b \ominus a)$, where $b \ominus a$ is an inverse to $\oplus$, i.e., the value for which $a \oplus (b \ominus a) = b$.

By definition of the new operation $\ominus$, we have

$$b = (a \oplus b) \ominus b.$$

Thus, we have

$$c(a, a \oplus b) = c_1((a \oplus b) \ominus a) = c_1(a, b).$$

In other words, for every a and b, we have

$$c_1(a, b) = c(a, a \oplus b).$$

Substituting this expression for $c_1(a, b)$ into the above formula, we conclude that

$$c(\mu_1, \mu_1 \oplus \mu_2 \oplus \mu_3) = c(\mu_1 \oplus \mu_2, \mu_1 \oplus \mu_2 \oplus \mu_3) \cdot c_1(\mu_1, \mu_1 \oplus \mu_2).$$

The left-hand side depends only on two values $x \stackrel{\text{def}}{=} \mu_1$ and $z \stackrel{\text{def}}{=} \mu_1 \oplus u_2 \oplus \mu_3$, and does not depend on the value $y \stackrel{\text{def}}{=} \mu_1 \oplus \mu_3$:

$$c(x, z) = c(y, z) \cdot c(x, y).$$

Thus, if we fix some value y_0, we conclude that

$$c(x, z) = g(z) \cdot h(x),$$

where we denoted $g(z) \stackrel{\text{def}}{=} c(y_0, z)$ and $h(x) \stackrel{\text{def}}{=} c(x, y_0)$.

Describing $c_1(a, b)$ in terms of the auxiliary function $c(a, b)$, we can transform the expression for the equivalent monetary value to

$$\sum_{i=1}^{n} c(\mu_i, \mu_1 \oplus \cdots \oplus \mu_n) \cdot x_i.$$

Substituting the expression $c(x, z) = g(z) \cdot h(x)$ into this formula, we conclude that the equivalent monetary value takes the form

$$\sum_{i=1}^{n} h(\mu_i) \cdot g \cdot x_i,$$

where we denoted $g \stackrel{\text{def}}{=} g(\mu_1 \oplus \cdots \oplus \mu_n)$.

For the case when $x_1 = x_2 = \cdots = x_n$, the boundedness requirement implies that the equivalent value is equal to x_1. Thus, we have

$$x_1 = \sum_{i=1}^{n} h(\mu_i) \cdot g \cdot x_1.$$

Dividing both sides by x_1, we conclude that

$$1 = g \cdot \sum_{i=1}^{n} h(\mu_i)$$

and hence, that

$$g = \frac{1}{\sum\limits_{i=1}^{n} h(\mu_i)}.$$

So, the equivalent monetary value is equal to the following expression:

$$\frac{\sum\limits_{i=1}^{n} h(\mu_i) \cdot x_i}{\sum\limits_{i=1}^{n} h(\mu_i)}.$$

So, now we are down to a single unknown function $h(\mu)$.

6.4.3 Conclusions

General conclusion. We need to decide between several actions. For each action, we know the possible outcomes $x_1, \ldots, x_n$, and for each of these possible outcomes i, we know the degree of possibility μ_i of this outcome. The above analysis shows that the benefit of each action can then be described by the following formula

$$\frac{\sum\limits_{i=1}^{n} h(\mu_i) \cdot x_i}{\sum\limits_{i=1}^{n} h(\mu_i)},$$

for an appropriate function $h(\mu)$.

How can we find the function $h(\mu)$**?** If we have two alternatives with the same outcome $x_1 = x_2$, then we can:

- either treat them separately, leading to the terms

$$h(\mu_1) \cdot g(\mu_1 \oplus \mu_2 \oplus \cdots) \cdot x_1 + h(\mu_2) \cdot g(\mu_1 \oplus \mu_2 \oplus \cdots) \cdot x_1 + \cdots$$

- or treat them as a single alternative x_1, with degree of possibility $\mu_1 \oplus \mu_2$, thus leading to the term

$$h(\mu_1 \oplus \mu_2) \cdot g(\mu_1 \oplus \mu_2 \oplus \cdots) \cdot x_1.$$

These two expressions must coincide, so we must have

$$h(\mu_1 \oplus \mu_2) = h(\mu_1) + h(\mu_2).$$

Let us show, on two specific cases, what this leads to.

Probabilistic case. In this case, the values μ_i are probabilities, and as we have mentioned, we have $\mu_1 \oplus \mu_2 = \mu_1 + \mu_2$. So, the above condition takes the form

$$h(\mu_1 + \mu_2) = h(\mu_1) + h(\mu_2).$$

Thus, in the probabilistic case, the function $h(\mu)$ must be additive.

The higher probability, the more importance should be given to the corresponding alternative, so the function $h(\mu)$ should be monotonic. It is known (see, e.g., [2]) that every monotonic additive function is linear, so we must have $h(\mu) = c \cdot \mu$ for some constant μ. Thus, the above formula for the equivalent amount takes the form

$$\frac{\sum_{i=1}^{n} c \cdot \mu_i \cdot x_i}{\sum_{i=1}^{n} c \cdot \mu_i}.$$

For probabilities, $\sum_{i=1}^{n} \mu_i = 1$. So, dividing both the numerator and the denominator by c, we conclude that the equivalent benefit has the form

$$\sum_{i=1}^{n} \mu_i \cdot x_i.$$

This is exactly the formula for the expected utility that appears when we consider the decision of rational agents under probabilistic uncertainty [19, 46, 59, 84].

Fuzzy case. In the fuzzy case, $a \oplus b$ is a t-conorm. It is known (see, e.g., [60]) that every t-conorm can be approximated, with arbitrary accuracy, by an Archimedean t-conorm, i.e., by a function of the type $G^{-1}(G(a) + G(b))$, where $G(a)$ is a strictly increasing continuous function and G^{-1} denotes the inverse function. Thus, from the practical viewpoint, we can safely assume that the actual t-conorm operation $a \oplus b$ is Archimedean:

$$a \oplus b = G^{-1}(G(a) + G(b)).$$

In this case, the condition $a \oplus b = c$ is equivalent to

$$G(a) + G(b) = G(c).$$

The requirement that

$$h(\mu_1 \oplus \mu_2) = h(\mu_1) + h(\mu_2)$$

means that if $a \oplus b = c$, then

$$h(a) + h(b) = h(c).$$

In other words, if $G(a) + G(b) = G(c)$, then

$$h(a) + h(b) = h(c).$$

If we denote $A \stackrel{\text{def}}{=} G(a)$, $B \stackrel{\text{def}}{=} G(b)$, and $C \stackrel{\text{def}}{=} h(c)$, then $a = G^{-1}(A)$, $b = G^{-1}(B)$, $c = G^{-1}(C)$, and the above requirement takes the following form: if $A + B = C$, then

$$h(G^{-1}(A)) + h(G^{-1}(B)) = h(G^{-1}(C)).$$

So, for the auxiliary function $H(A) \stackrel{\text{def}}{=} h(G^{-1}(A))$, we have $A + B = c$ implying that $H(C) = H(A) + H(B)$, i.e., that $H(A + B) = H(A) + H(B)$. The function $H(A)$ is monotonic and additive, hence $H(A) = k \cdot A$ for some constant k.

So, $H(A) = h(G^{-1}(A)) = k \cdot A$. Substituting $A = G(a)$ into this formula, we conclude that

$$h(G^{-1}(G(a))) = h(a) = k \cdot G(a).$$

Thus, in the fuzzy case, the equivalent monetary value of each action is equal to

$$\frac{\sum\limits_{i=1}^{n} k \cdot G(\mu_i) \cdot x_i}{\sum\limits_{i=1}^{n} k \cdot G(\mu_i)}.$$

Dividing both the numerator and the denominator by the constant k, we get the final formula

$$\frac{\sum_{i=1}^{n} G(\mu_i) \cdot x_i}{\sum_{i=1}^{n} G(\mu_i)},$$

where $G(a)$ is a "generating" function of the t-conorm, i.e., a function for which the t-conorm has the form

$$G^{-1}(G(a) + G(b)).$$

Fuzzy case: example. For example, for the algebraic sum t-conorm

$$a \oplus b = a + b - a \cdot b,$$

we have

$$1 - a \oplus b = (1 - a) \cdot (1 - b)$$

and thus,

$$- \ln(1 - a \oplus b) = (- \ln(1 - a)) + (- \ln(1 - b)),$$

so we have $G(a) = - \ln(1 - a)$.

Thus, the formula for the equivalent amount takes the form

$$\frac{\sum_{i=1}^{n} \ln(1 - \mu_i) \cdot x_i}{\sum_{i=1}^{n} \ln(1 - \mu_i)}.$$

Chapter 7
Conclusions

In many practical application, we process measurement results and expert estimates. Measurements and expert estimates are never absolutely accurate, their results are slightly different from the actual (unknown) values of the corresponding quantities. It is therefore desirable to analyze how this measurement and estimation inaccuracy affects the results of data processing.

There exist numerous methods for estimating the accuracy of the results of data processing under different models of measurement and estimation inaccuracies: probabilistic, interval, and fuzzy. To be useful in engineering applications, these methods:

- should provide accurate estimate for the resulting uncertainty,
- should not take too much computation time,
- should be understandable to engineers, and
- should be sufficiently general to cover all kinds of uncertainty.

In this book, on several case studies, we show how we can achieve these four objectives:

- We show that we can get more accurate estimates by properly taking model inaccuracy into account.
- We show that we can speed up computations by processing different types of uncertainty differently.
- We show that we can make uncertainty-estimating algorithms more understandable.
- We also analyze how general uncertainty-estimating algorithms can be.

We also analyze how to make decisions under different types of uncertainty.

We apply the corresponding algorithms to practical engineering problems, in particular, to the inverse problem in geosciences and to the problem of pavement compaction.

© Springer International Publishing AG, part of Springer Nature 2018
A. Pownuk and V. Kreinovich, *Combining Interval, Probabilistic, and Other Types of Uncertainty in Engineering Applications*, Studies in Computational Intelligence 773, https://doi.org/10.1007/978-3-319-91026-0_7

References

1. J. Aczél, *Lectures on Functional Equations and Their Applications* (Dover, New York, 2006)
2. J. Aczél, H. Dhombres, *Functional Equations in Several Variables* (Cambridge University Press, Cambridge, 1989)
3. P.A. Arkhipov, S.I. Kumkov et al., Estimation of Pb activity in double systems Ph-Sb and Pb-Bi. Rasplavy **5**, 43–52 (2012) (in Russian)
4. M.G. Averill, Lithospheric investigation of the Southern Rio Grande Rift, University of Texas at El Paso, Department of Geological Sciences, Ph.D. Dissertation (2007)
5. P. Barragan, S. Nazarian, V. Kreinovich, A. Gholamy, M. Mazari, How to estimate resilient modulus for unbound aggregate materials: a theoretical explanation of an empirical formula, in *Proceedings of the 2016 World Conference on Soft Computing, Berkeley, California, 22–25 May 2016*, pp. 203–207
6. E. Bishop, D. Bridges, *Constructive Analysis* (Springer, Heidelberg, 1985)
7. R. Bonola, *Non-Euclidean Geometry* (Dover Publications, New York, 2007)
8. B.G. Buchanan, E.H. Shortliffe, *Rule Based Expert Systems: The MYCIN Experiments of the Stanford Heuristic Programming Project* (Addison-Wesley, Reading, 1984)
9. R.L. Burden, J.D. Faires, A.M. Burden, *Numerical Analysis* (Cengage Learning, Boston, 2015)
10. D.G. Cacuci, *Sensitivity and Uncertainty Analysis: Theory* (Chapman & Hall/CRC, Boca Raton, 2007)
11. S. Chakraverty, M. Hladík, D. Behera, Formal solution of an interval system of linear equations with an application in static responses of structures with interval forces. Appl. Math. Model. **50**, 105–117 (2017)
12. S. Chakraverty, M. Hladík, N.R. Mahato, A sign function approach to solve algebraically interval system of linear equations for nonnegative solutions. Fundamenta Informaticae **152**, 13–31 (2017)
13. S. Dick, Toward complex fuzzy logic. IEEE Trans. Fuzzy Syst. **13**(3), 405–414 (2005)
14. P.A.M. Dirac, The physical interpretation of quantum mechanics. Proc. R. Soc. A: Math. Phys. Eng. Sci. **180**(980), 1–39 (1942)
15. L. Dymova, P. Sevastjanov, A. Pownuk, V. Kreinovich, Practical need for algebraic (equality-type) solutions of interval equations and for extended-zero solutions, in *Proceedings of the 12th International Conference on Parallel Processing and Applied Mathematics PPAM'17, Lublin, Poland, 10–13 September 2017* (to appear)
16. H.B. Enderton, *A Mathematical Introduction to Logic* (Academic Press, New York, 2001)

© Springer International Publishing AG, part of Springer Nature 2018
A. Pownuk and V. Kreinovich, *Combining Interval, Probabilistic, and Other Types of Uncertainty in Engineering Applications*, Studies in Computational Intelligence 773, https://doi.org/10.1007/978-3-319-91026-0

17. R.P. Feynman, Negative probability, in *Quantum Implications: Essays in Honour of David Bohm*, ed. by F.D. Peat, B. Hiley (Routledge & Kegan Paul Ltd., Abingdon-on-Thames, 1987), pp. 235–248
18. R. Feynman, R. Leighton, M. Sands, *The Feynman Lectures on Physics* (Addison Wesley, Boston, 2005)
19. P.C. Fishburn, *Utility Theory for Decision Making* (Wiley, New York, 1969)
20. R. Floss, G, Bräu, M. Gahbauer, N. Griber, J. Obermayer, Dynamische Vedichtungsprüfung bei Erd-und Straß enbauten (Dynamic Compaction Testing in Earth and Road Construction), Prümaft für Grundbau, Boden- und Felsmechanik, Technische Universerität München, Heft 612, München (1991)
21. L. Forssblad, Compaction meter on vibrating rollers for improved compaction control, in *Proceedings of the International Conference on Compaction, Paris, France*, vol. II (1980), pp. 541–546
22. S.V. Glakovsky, S.I. Kumkov, Application of approximation methods to analysis of peculiarities of breaking-up and to forecasting break-resistibility of high-strength steel, in *Mathematical Modeling of Systems and Processes*, Proceedings of the Perm State Technical University, vol. 5 (1997), pp. 26–34 (in Russian)
23. C.A. Hales, S.A. Stuart, M.H. Anderson, E.S.J. Robinson, Modelling cognitive affective biases in major depressive disorder using rodents. Br. J. Pharmacol. **171**(20), 4524–4538 (2014)
24. J.A. Hole, Nonlinear high-resolution three-dimensional seismic travel time tomography. J. Geophys. Res. **97**(B5), 6553–6562 (1992)
25. L. Hurwicz, Optimality criteria for decision making under ignorance, Cowles Commission Discussion Paper, Statistics, No. 370 (1951)
26. L. Jaulin, M. Kiefer, O. Dicrit, E. Walter, *Applied Interval Analysis* (Springer, London, 2001)
27. C. Kiekintveld, M.T. Islam, V. Kreinovich, Security fame with interval uncertainty, in *Proceedings of the Twelfth International Conference on Autonomous Agents and Multiagent Systems AAMAS'2013, Saint Paul, Minnesota, 6–10 May 2013*, ed. by T. Ito, C. Jonker, M. Gini, O. Shehory, pp. 231–238
28. G. Klir, B. Yuan, *Fuzzy Sets and Fuzzy Logic* (Prentice Hall, Upper Saddle River, 1995)
29. V. Kreinovich, Error estimation for indirect measurements is exponentially hard. Neural, Parallel Sci. Comput. **2**(2), 225–234 (1994)
30. V. Kreinovich, Relation between interval computing and soft computing, in *Knowledge Processing with Interval and Soft Computing*, ed. by C. Hu, R.B. Kearfott, A. de Korvin, V. Kreinovich (Springer, London, 2008), pp. 75–97
31. V. Kreinovich, Interval computations and interval-related statistical techniques: tools for estimating uncertainty of the results of data processing and indirect measurements, in *Data Modeling for Metrology and Testing in Measurement Science*, ed. by F. Pavese, A.B. Forbes (Birkhauser-Springer, Boston, 2009), pp. 117–145
32. V. Kreinovich, Decision making under interval uncertainty (and beyond), in *Human-Centric Decision-Making Models for Social Sciences*, ed. by P. Guo, W. Pedrycz (Springer, Berlin, 2014), pp. 163–193
33. V. Kreinovich, J. Beck, C. Ferregut, A. Sanchez, G.R. Keller, M. Averill, S.A. Starks, Monte-Carlo-type techniques for processing interval uncertainty, and their potential engineering applications. Reliab. Comput. **13**(1), 25–69 (2007)
34. V. Kreinovich, S. Ferson, A new Cauchy-based black-box technique for uncertainty in risk analysis. Reliab. Eng. Syst. Saf. **85**(1–3), 267–279 (2004)
35. V. Kreinovich, O. Kosheleva, A. Pownuk, R. Romero, How to take into account model inaccuracy when estimating the uncertainty of the result of data processing, in *Proceedings of the ASME 2015 International Mechanical Engineering Congress and Exposition IMECE'2015, Houston, Texas, 13–19 November 2015*
36. V. Kreinovich, O. Kosheleva, A. Pownuk, R. Romero, How to take into account model inaccuracy when estimating the uncertainty of the result of data processing. ASCE-ASME J. Risk Uncertain. Eng. Syst. Part B: Mech. Eng. **3**(1), Paper No. 011002 (2017)

37. V. Kreinovich, A. Lakeyev, J. Rohn, P. Kahl, *Computational Complexity and Feasibility of Data Processing and Interval Computations* (Kluwer, Dordrecht, 1997)
38. V. Kreinovich, A. Pownuk, O. Kosheleva, Combining interval and probabilistic uncertainty: what is computable?, in *Advances in Stochastic and Deterministic Global Optimization*, ed. by P. Pardalos, A. Zhigljavsky, J. Zilinskas (Springer, Cham, 2016), pp. 13–32
39. B.J. Kubica, A. Pownuk, V. Kreinovich, What decision to make in a conflict situation under interval uncertainty: efficient algorithms for the Hurwicz approach, in *Proceedings of the 12th International Conference on Parallel Processing and Applied Mathematics PPAM'17, Lublin, Poland, 10–13 September 2017* (to appear)
40. S.I. Kumkov, Processing the experimental data on ion conductivity of molten electrolyte by the interval analysis method. Rasplavy **3**, 86–96 (2010) (in Russian)
41. S.I. Kumkov, Estimation of spring stiffness under conditions of uncertainty: interval approach, in *Proceedings of the 1st International Workshop on Radio Electronics and Information Technologies REIT'2017, Ekaterinburg, Russia, 15 March 2017* (in Russian)
42. S.I. Kumkov, V. Kreinovich, A. Pownuk, In system identification, interval (and fuzzy) estimates can lead to much better accuracy than the traditional statistical ones: general algorithm and case study, in *Proceedings of the IEEE Conference on Systems, Man, and Cybernetics SMC'2017, Banff, Canada, 5–8 October 2017* (to appear)
43. S.I. Kumkov, Yu.V. Mikushina, Interval approach to identification of catalytical process parameters. Reliab. Comput. **19**, 197–214 (2013)
44. K. Kunen, *Set Theory* (College Publications, London, 2011)
45. A. Lakeyev, On the computational complexity of the solution of linear systems with moduli. Reliab. Comput. **2**(2), 125–131 (1996)
46. R.D. Luce, R. Raiffa, *Games and Decisions: Introduction and Critical Survey* (Dover, New York, 1989)
47. Y. Lucet, Faster than the fast Legendre transform, the linear-time Legendre transform. Numer. Algorithms **16**, 171–185 (1997)
48. D.G. Luenberger, Y. Ye, *Linear and Nonlinear Programming* (Springer, Cham, 2016)
49. M. Mazari, F. Navarro, I. Abdallah, S. Nazarian, Comparison of numerical and experimental responses of pavement systems using various resilient modulus models. Soils Found. **54**(1), 36–44 (2014)
50. P. Melin, O. Castillo, A. Pownuk, O. Kosheleva, V. Kreinovich, How to gauge the accuracy of fuzzy control recommendations: a simple idea, in *Proceedings of the 2017 Annual Conference of the North American Fuzzy Information Processing Society NAFIPS'2017, Cancun, Mexico, 16–18 October 2017* (to appear)
51. J.M. Mendel, *Uncertain Rule-Based Fuzzy Logic Systems: Introduction and New Directions* (Prentice-Hall, Upper Saddle River, 2001)
52. J.M. Mendel, D. Wu, *Perceptual Computing: Aiding People in Making Subjective Judgments* (IEEE Press and Wiley, New York, 2010)
53. G.A. Miller, The magical number seven, plus or minus two: some limits on our capacity for processing information. Psychol. Rev. **63**(2), 81–97 (1956)
54. M.A. Mooney, D. Adam, Vibratory roller integrated measurement of earthwork compaction: an overview, in *Proceedings of the International Symposium on Field Measurements in Geomechanics FMGM'2007, Boston, Massachusetts, 24–27 September 2007*
55. M.A. Mooney, P.B. Gorman, E. Farouk, J.N. Gonzalez, A.S. Akanda, Exploring vibration-based intelligent soft compaction, Oklahoma Department of Transportation, Projet No. 2146, Final Report (2003)
56. M.A. Mooney, P.B. Gorman, J.N. Gonzalez, Vibration-based health monitoring during eathwork construction. J. Struct. Health Monit. **2**(4), 137–152 (2005)
57. M.A. Mooney, R.V. Rinehart, N.W. Facas, O.M. Musimbi, D.J. White, Intelligent Soil Compaction Systems, National Cooperative Highway Research Program (NCHRP) Report 676, Transportation Research Board, Washington, DC (2010)
58. R.E. Moore, R.B. Kearfott, M.J. Cloud, *Introduction to Interval Analysis* (SIAM, Philadelphia, 2009)

59. H.T. Nguyen, O. Kosheleva, V. Kreinovich, Decision making beyond arrow's 'impossibility theorem', with the analysis of effects of collusion and mutual attraction. Int. J. Intell. Syst. **24**(1), 27–47 (2009)
60. H.T. Nguyen, V. Kreinovich, P. Wojciechowski, Strict Archimedean t-norms and t-conorms as universal approximators. Int. J. Approx. Reason. **18**(3–4), 239–249 (1998)
61. H.T. Nguyen, E.A. Walker, *A First Course in Fuzzy Logic* (Chapman and Hall/CRC, Boca Raton, 2006)
62. K. Nickel, Die Auflösbarkeit linearer Kreisscheineb- und Intervall-Gleichingssyteme. Linear Algebra Appl. **44**, 19–40 (1982)
63. M.A. Nielsen, I.L. Chuang, *Quantum Computation and Quantum Information* (Cambridge University Press, Cambridge, 2000)
64. P.S.K. Ooi, A.R. Archilla, K.G. Sandefur, Resilient modulus models for comactive cohesive soils. Transp. Res. Record **1874**, 115–124 (2006)
65. J. Panksepp, J.S. Wright, M.D. Döbrössy, Th.E. Schlaepfer, V.A. Coenen, Affective neuroscience strategies for understanding and treating depression: from preclinical models to three novel therapeutics. Clin. Psychol. Sci. **2**(4), 472–494 (2014)
66. J. Papciak, P. Popik, E. Fuchs, R. Rygula, Chronic psychosocial stress makes rats more 'pessimistic' in the ambiguous-cue interpretation paradigm. Behav. Brain Res. **256**, 305–310 (2013)
67. P. Pinheiro da Silva, A. Velasco, M. Ceberio, C. Servin, M.G. Averill, N. Del Rio, L. Longpré, V. Kreinovich, Propagation and provenance of probabilistic and interval uncertainty in cyberinfrastructure-related data processing and data fusion, in *Proceedings of the International Workshop on Reliable Engineering Computing REC'08, Savannah, Georgia, 20–22 February 2008*, ed. by R.L. Muhanna, R.L. Mullen, pp. 199–234
68. A.M. Potapov, S.I. Kumkov, Y. Sato, Processing of experimental data on viscosity under one-sided character of measuring errors. Rasplavy **3**, 55–70 (2010) (in Russian)
69. A. Pownuk, Approximate method for computing the sum of independent random variables, in *Abstracts of the 17th Joint UTEP/NMSU Workshop on Mathematics, Computer Science, and Computational Sciences, El Paso, Texas, 7 November 2015*
70. A. Pownuk, P. Barragan Olague, V. Kreinovich, Why compaction meter value (CMV) is a good measure of pavement stiffness: towards a possible theoretical explanation. Math. Struct. Model. **40**, 48–54 (2016)
71. A. Pownuk, O. Kosheleva, V. Kreinovich, Towards decision making under general uncertainty. Math. Struct. Model. **44** (2017) (to appear)
72. A. Pownuk, V. Kreinovich, Which point from an interval should we choose?, in *Proceedings of the 2016 Annual Conference of the North American Fuzzy Information Processing Society NAFIPS'2016, El Paso, Texas, October 31–November 4 2016*
73. A. Pownuk, V. Kreinovich, Isn't every sufficiently complex logic multi-valued already: Lindenbaum-Tarski algebra and fuzzy logic are both particular cases of the same idea, in *Proceedings of the Joint 17th Congress of International Fuzzy Systems Association and 9th International Conference on Soft Computing and Intelligent Systems IFSA-SCIS'2017, Otsu, Japan, 27–30 June 2017*
74. A. Pownuk, V. Kreinovich, (Hypothetical) negative probabilities can speed up uncertainty propagation algorithms, in *Quantum Computing: an Environment for Intelligent Large Scale Real Application*, ed. by A.E. Hassanien, M. Elhoseny, A. Farouk, J. Kacprzyk (Springer) (to appear)
75. A. Pownuk, V. Kreinovich, Why mixture of probability distributions? Int. J. Intell. Technol. Appl. Stat. (IJITAS) (to appear)
76. A. Pownuk, V. Kreinovich, Why linear interpolation? Math. Struct. Model. **43** (2017) (to appear)
77. A. Pownuk, V. Kreinovich, Towards decision making under interval uncertainty. J. Uncertain. Syst. (to appear)

78. A. Pownuk, V. Kreinovich, Why unexpectedly positive experiences make decision makers more optimistic: an explanation, in *Proceedings of the 10th International Workshop on Constraint Programming and Decision Making CoProd'2017, El Paso, Texas, 3 November 2017* (to appear)

79. A. Pownuk, V. Kreinovich, Which value x best represents a sample $x_1, \ldots, x_n$: utility-based approach under interval uncertainty, in *Proceedings of the 10th International Workshop on Constraint Programming and Decision Making CoProd'2017, El Paso, Texas, 3 November 2017* (to appear)

80. A. Pownuk, V. Kreinovich, Decomposition into granules speeds up data processing under uncertainty, University of Texas at El Paso, Department of Computer Science, Technical Report UTEP-CS-17-72 (2017)

81. A. Pownuk, V. Kreinovich, S. Sriboonchitta, Fuzzy data processing beyond min t-norm, in *Complex Systems: Solutions and Challenges in Economics, Management, and Engineering*, ed. by C. Berger-Vachon, A.M. Gil Lafuente, J. Kacprzyk, Y. Kondratenko, J.M. Merigo Lindahl, C. Morabito (Springer) (to appear)

82. S.G. Rabinovich, *Measurement Errors and Uncertainty. Theory and Practice* (Springer, Berlin, 2005)

83. K. Ratschek, W. Sauer, Linear interval equations. Computing **25**, 105–115 (1982)

84. H. Raiffa, *Decision Analysis* (Addison-Wesley, Reading, 1970)

85. A.A. Redkin, Yu.P. Zalkov, I.V. Korzun, O.G. Reznitskih, T.V. Yaroslavela, S.I. Kumkov, Heat capacity of molten halides. J. Phys. Chem. **119**, 509–512 (2015)

86. S.K. Reed, *Cognition: Theories and Application* (Wadsworth Cengage Learning, Belmont, 2010)

87. R.T. Rockafeller, *Convex Analysis* (Princeton University Press, Princeton, 1997)

88. R. Rygula, H. Pluta, P. Popik, Laughing rats are optimistic. PLoS ONE **7**(2) (2012), Paper e51959

89. A. Saltelli, K. Chan, E.M. Scott, *Sensitivity Analysis* (Wiley, Chichester, 2009)

90. G. Samorodnitsky, M.S. Taqqu, *Stable Non-Gaussian Random Processes: Stochastic Models with Infinite Variance* (Chapman & Hall, New York, 1994)

91. J.R. Schoenfeld, *Mathematical Logic* (Addison-Wesley, Reading, 2001)

92. P. Sevastjanov, L. Dymova, Fuzzy solution of interval linear equations, in *Proceedings of the 7th International Conference on Parallel Processing and Applied Mathematics PPAM'2017, Gdansk, Poland* (2007), pp. 1392–1399

93. P. Sevastjanov, L. Dymova, A new method for solving interval and fuzzy equations: linear case. Inf. Sci. **17**, 925–937 (2009)

94. S.P. Shary, Algebraic approach to the interval linear static identification, tolerance, and control problems, or one more application of Kaucher arithmetic. Reliab. Comput. **2**(1), 3–33 (1996)

95. S.P. Shary, Algebraic approach in the 'outer problem' for interval linear equations. Reliab. Comput. **3**(2), 103–135 (1997)

96. S.P. Shary, A new technique in systems analysis under interval uncertainty and ambiguity. Reliab. Comput. **8**, 321–418 (2002)

97. D.J. Sheskin, *Handbook of Parametric and Nonparametric Statistical Procedures* (Chapman and Hall/CRC, Boca Raton, 2011)

98. M. Sipser, *Introduction to Theory of Computation* (Thomson Course Technology, Boston, 2012)

99. P. Solin, K. Segeth, I. Dolezel, *Higher-Order Finite Element Methods* (Chapman & Hall/CRC, Boca Raton, 2003)

100. C.D. Stylios, A. Pownuk, V. Kreinovich, Sometimes, it is beneficial to process different types of uncertainty separately, in *Proceedings of the Annual Conference of the North American Fuzzy Information Processing Society NAFIPS'2015 and 5th World Conference on Soft Computing, Redmond, Washington, 17–19 August 2015*

101. S. Tadelis, *Game Theory: an Introduction* (Princeton University Press, Princeton, 2013)

102. H.F. Thurner, L. Forsblad, Compaction meter on vibrating rollers, Research Bulletin No. 8022, Solna, Sweden

103. H.F. Thurner, A. Sandström, A new device for instant compaction control, in *Proceedings of International Conference on Compaction, Paris, France*, vol. II (1980), pp. 611–614
104. G.W. Walster, V. Kreinovich, For unknown-but-bounded errors, interval estimates are often better than averaging. ACM SIGNUM Newsl. **31**(2), 6–19 (1996)
105. K. Weihrauch, *Computable Analysis* (Springer, Berlin, 2000)
106. D. White, M. Thompson, Relationships between in-situ and roller-integrated compaction measurements for granular soils. ASCE J. Geotech. Geomech. Eng. **134**(12), 1763–1770 (2008)
107. D. White, M. Thompson, P. Vennapusa, Filed validation of intelligent compaction monitoring technology for unbound materials, Report No. MN/RC 2007-10, Minnesota Department of Transportation, St. Paul, Minnesota (2008)
108. L.A. Zadeh, Fuzzy sets. Inf. Control **8**, 338–353 (1965)

Index

A
Additive function, 188
Additivity, 182
Alpha-cut, 10, 46, 58
Analytical differentiation, 4
"And"-operation, 10, 49, 52
Applications to
    animal behavior, 177
    catalysis, 24
    energy storage, 23
    engineering, 11
    geophysics, 36, 52
    mechanics, 24
    medicine, 52
    transportation engineering, 132
Axiom, 123
    of Choice, 123

B
Battery, 23
    liquid metal, 23
    molten salt, 23
Best-case situation, 159, 166, 167, 178
Bisection, 7
Box, 101

C
Catalysis, 24
Cauchy distribution, 6
Cdf, 141
Central Limit Theorem, 13, 15, 34, 107

Centroid defuzzification, 42
Characteristic function, 65, 70, 71, 80, 88, 114
Compaction Meter Value (CMV), 132
Compactor
    rolling, 132
    vibrating, 132
Compact set, 147
Computable, 137
    cdf, 144
    compact, 147
    function, 139
    metric space, 148
    probability distribution, 148
    real number, 139
    set, 147
Computation time, 4, 29, 57–59, 61, 65, 66, 85, 86, 171
Confidence interval, 35
Conflict situation, 167
Conservativeness, 126
Consistency, 125
Continuum Hypothesis, 123
Control
    of an airplane, 1
    solution, 99
Convex analysis, 60
Correlation, 137
Cumulant, 80
Cumulative distribution function, 141

D
Data processing, 1

© Springer International Publishing AG, part of Springer Nature 2018   199
A. Pownuk and V. Kreinovich, *Combining Interval, Probabilistic, and Other Types of Uncertainty in Engineering Applications*, Studies in Computational Intelligence 773, https://doi.org/10.1007/978-3-319-91026-0

Decision making, 12, 26, 27, 80, 142, 157,
        158, 177, 181
    fuzzy case, 188
    interval case, 27, 177
    probabilistic case, 176, 188
Decision theory, 26, 27, 80, 142, 158, 177
Defuzzification, 42
Differentiation
    analytical, 4
    numerical, 4, 6
Discretization, 32
Distribution
    Cauchy, 6, 32, 36, 65, 93, 114
    Gaussian, 5, 13, 15, 34, 65, 71, 86, 107,
        142, 143
    infinitely divisible, 65
    Laplace, 71
    normal, 5, 13, 15, 34, 65, 71, 86, 107,
        142, 143
    uniform, 7, 142
Disutility, 26

E
Electric car, 23
Ellipsoid, 101
Energy
    solar, 23
    wind, 23
Energy density, 23
Energy storage, 23
Epi-sum, 60
Epsilon-net, 149
Estimation, 1
    accurate, 11
    fast, 11
    general, 12
    understandable, 12
Euclid's Vth Postulate, 122
Excess, 81
Extended zero, 106

F
Fast Fourier Transform, 65, 70, 132
Fast Legendre Transform, 62
Finite Element Method, 32, 38
Frequency, 143
Functional equation, 89
Fusion heat, 23
Fuzzy
    control, 40
    elicitation, 41
    interval-valued, 42, 181

    number, 10
    type-2, 42, 181
    uncertainty, 3, 10, 45, 49, 119, 188

G
Gaussian distribution, 13, 15, 34, 86, 107
Geometry
    Euclid's Vth Postulate, 122
    Lobachevsky, 122
    non-Euclidean, 122
Geophysics, 36
Gödel's theorem, 121
Granular computing, 63
Granularity, 63
Granule, 63

H
High-performance computers, 1
Hole's algorithm, 36
Hurwicz criterion, 27, 159, 166, 167, 172,
        177

I
Inconsistent constraints, 19
Independent quantities
    interval case, 102
    probabilistic case, 5, 64, 89
Infimal convolution, 60
Inflation, 104
Intelligent compaction, 132
Interpolation, 125
    linear, 125
Interval, 2, 15
    computations, 3, 6, 15, 58, 99, 101
    uncertainty, 2, 15, 25, 27, 47, 97, 137,
        157, 159, 171
    -valued fuzzy, 42, 181
Inverse Fourier transform, 65, 91, 94

L
Least Squares method, 17, 67
Legendre–Fenchel transform, 60
Legendre transform, 60
    Fast, 62
Lindenbaum–Tarski algebra, 122
Linear
    interpolation, 125
    model, 125
    order, 123
    programming, 18, 170
    -time algorithm, 61

Linearization, 3, 5, 6, 10, 17, 54, 63, 82
Liquid metal battery, 23
Logic
    2-valued, 119
    fuzzy, 119, 121
    multi-valued, 119

M
Magical number seven ± two, 11
Maximum likelihood method, 7
Mean, 137
Measurable function, 90
Measurement, 1, 2
    error, 2
    uncertainty, 2, 15
Measuring instrument, 2
Mechanics, 24
Membership function, 3, 40
    non-normalized, 41
    normalization, 41
    normalized, 41, 61
    triangular, 46
Meteorology, 1
Mixture, 115, 146
Model inaccuracy, 30
Molten salt battery, 23
Moment of a distribution, 80
Monte-Carlo method, 5, 79, 86, 94, 95
    for intervals, 6, 106
    realistic, 106
MYCIN, 52

N
Negative probability, 76
Newton's equations, 133
Non-linear model, 132
Non-normalized membership function, 41
Normal distribution, 5, 13, 15, 34, 86, 107
Normalization, 41
Normalized membership function, 41
NP-hard, 137
Numerical differentiation, 4, 6, 85

O
Octave, 75
Optimism, 176, 178
Optimism-pessimism parameter, 27, 159, 167, 172, 177
Order
    linear, 123
Oriental musical traditions, 75

"'Or"-operation, 10, 52, 184

P
Parallelization, 171
Parceval theorem, 67
Pavement, 132
Pdf, 64
Prediction, 1
Probabilistic uncertainty, 3, 25, 63, 115, 137, 176, 188
Probability
    amplitude, 76
    density function, 6, 40, 64, 76, 115, 141
    negative, 76

Q
Quadratic-time algorithm, 29
Quantum physics, 76

R
Random number generator, 7
Rat, 177
Roller, 132

S
Scale-invariance, 104, 125, 127
Science
    main objective, 1
Seismic inverse problem, 36
Sensitivity analysis
    global, 10
    local, 10, 32
Set theory, 123
Shift-invariance, 104
Skewness, 81
Solar energy, 23
Solution
    algebraic, 97, 103
    control, 97
    equality-type, 97, 103
    extended-zero, 97, 106
    formal, 103
    tolerance, 97
    united, 97, 100
Standard deviation, 40, 137
Step function, 141
    is not computable, 141
Stiffness, 132
Subjective probability, 178
System identification, 13, 15
    interval case, 16

probabilistic case, 17

T
T-conorm, 52, 184, 188
    algebraic sum, 52, 190
    Archimedean, 52, 188
    max, 52, 53
Tickling, 177
T-norm, 49, 52
    algebraic product, 52, 59
    Archimedean, 52, 59
    min, 49, 52, 57
Tolerance solution, 101
Type-2 fuzzy logic, 42, 181

U
Uncertainty
    combined, 137
    fuzzy, 3, 45, 49, 119, 181, 188
    fuzzy type-2, 181
    general, 180
    interval, 2, 25, 27, 47, 97, 137, 157, 159,
        171, 181

    possibilistic, 181
    probabilistic, 3, 25, 63, 115, 137, 176,
        181, 188
    underestimated, 19, 104
Underestimated uncertainty, 19, 104
Uniform distribution, 7
Universal approximator, 59
Utility, 25, 26, 80, 142, 158, 177

V
Variance, 5, 81, 86, 142
Vibration, 132

W
Wave function, 76
Western musical tradition, 75
Wind energy, 23
Worst-case situation, 159, 166, 167, 171, 178

Z
Zadeh's extension principle, 10, 49, 54
Zero-sum game, 168, 171, 175